The Noma Guide to
Fermentation

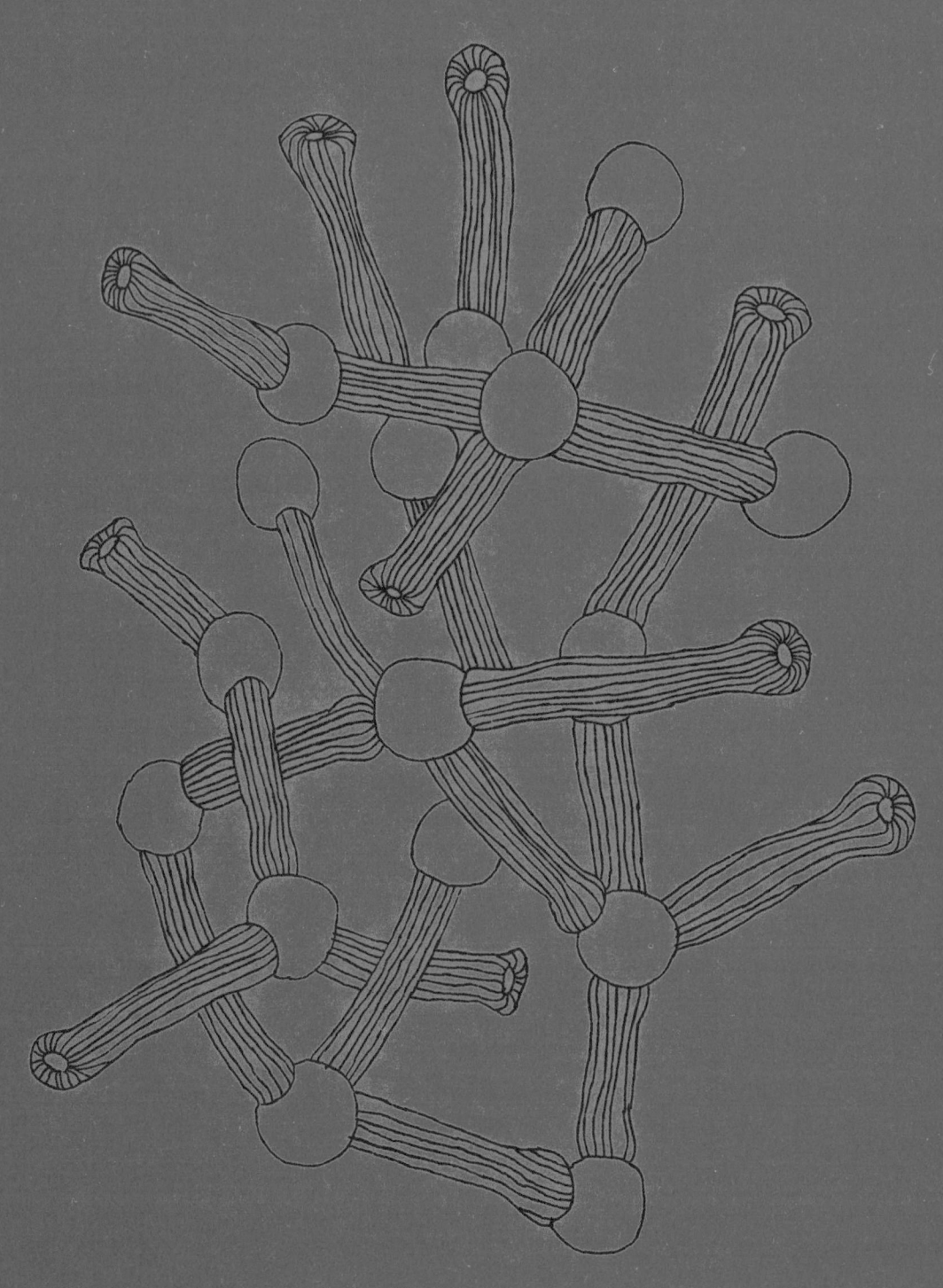

Foundations
of Flavor

The Noma Guide to Fermentation

René Redzepi & David Zilber

Photographs
by Evan Sung

Illustrations
by Paula Troxler

Artisan | New York

This book would not have been possible without the countless chefs and enthusiasts who have taken part in our never-ending quest of discovery. So many people have contributed small pieces to the great puzzle that has made Noma's world of fermentation what it is today. Notably, Dr. Arielle Johnson, Torsten Vildgaard, Lars Williams, Thomas Frebel, Rosio Sanchez, Josh Evans, Ben Reade, Roberto Flore, and all those involved in the Nordic Food Lab. If we have seen further, it is by standing on the shoulders of giants.

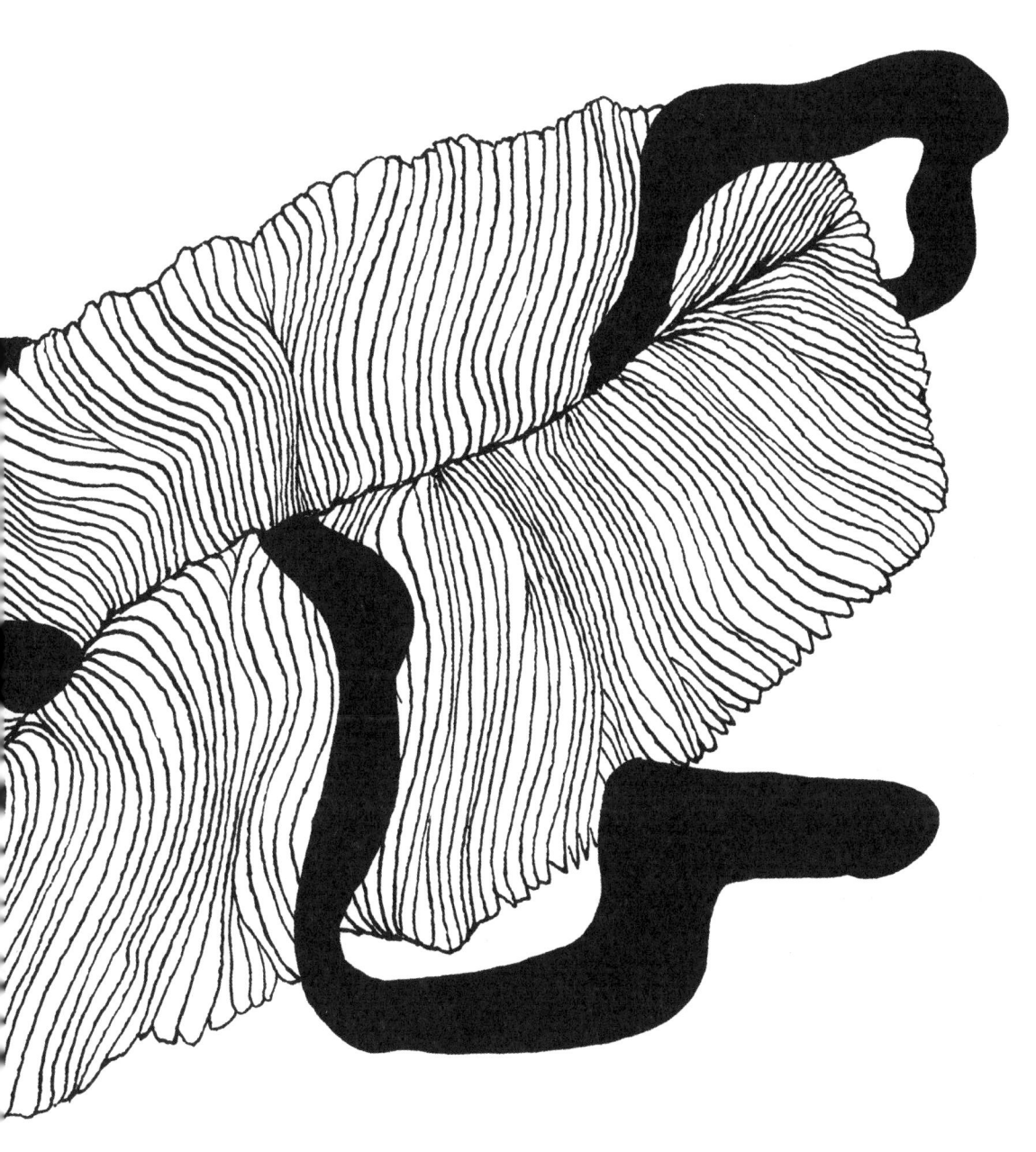

Introduction 9
About This Book 19

Primer 25
Lacto-Fermented Fruits
  and Vegetables 55
Kombucha 109
Vinegar 157
Koji 211
Misos and Peaso 269
Shoyu 329
Garum 361
Black Fruits and
  Vegetables 403

Equipment 442
Sources 448
Acknowledgments 449
Index 450

Noma in its new home on the outskirts
of the Christiania neighborhood in
Copenhagen. Opening week, February 2018.

Introduction
René Redzepi

Our story with fermentation is a story of accidents.

In the very early years of Noma, we were caught up in a search for ingredients, looking to stock our larder with things that could keep our cooking interesting through the colder months of the year.

I remember one day in the early summer when our longtime forager, Roland Rittman, walked through the door with a handful of odd little flower buds, round but also somehow triangular, perfectly juicy, with a flavor like ramps—not garlicky, exactly, but with that same punch and depth. We'd never tasted anything like it. Roland mentioned that these ramson "berries" used to be quite common in Nordic cuisine, and that people would preserve them for use through the winter.

And so we set out to make our own caper-like pickle of ramson buds. If you'd asked us what we thought was happening to the tiny garlicky orbs as they sat in a jar packed with salt, we would have described it as "curing" or "maturing." If you'd mentioned the concept of lactic acid fermentation, we would have cocked our heads and looked at you quizzically.

The ramson capers were a revelation. Suddenly we had this ingredient at our disposal that could bring little bursts of acidity and saltiness and pungency to any dish. And we didn't have to import it from somewhere else. It had grown in our own backyard and become something more, merely through the addition of salt.

One accidental success led to another.

I can't remember whose idea it was to salt gooseberries, but it was around 2008, so it must have been Torsten Vildgaard or Søren Westh. They were messing around with all kinds of things on the boat that was anchored in front of the restaurant.

No larger than a fishing vessel you might take out for a day on the ocean, the boat housed something we called the Nordic Food Lab. Its purpose was to investigate what could be done with the food in our region and share that knowledge freely with anyone interested. It was a place for long-term investigation, rather than a test kitchen for tinkering with next week's dishes. One of our chefs, Ben Reade, used to sleep among the ferments on that boat—that's the sort of character we had working in the lab.

One day, Torsten put a spoon in front of me with a slice of gooseberry that had been salted, vacuum-bagged, and fermented, then forgotten for a year. I tasted it and I was completely shocked. I know that probably sounds like an exaggeration—after all, we're talking about a spoonful of pickled berry. But you have to try to put yourself in my frame of mind: You've grown up in Scandinavia eating gooseberries your whole life, and now there's this thing in front of you. It tastes familiar but also like nothing you've ever had before, like an old comfortable sweater with bright new colors woven through the original fabric.

Today when I taste a pickled gooseberry, I recognize the unmistakable effect of lacto-fermentation, but that first time really changed everything for me and Noma. It was the beginning of a decade in which we would study fermentation with intense focus and enthusiasm.

—

I've forgotten so many details. I regret not taking more notes in those early days. Every week held a revelation of some sort, reached by the same basic train of thought: *We need more things to cook with. We have these seasonal ingredients. What can we do to make them better? What can we do to make them last?* At first, we had no idea how fermentation worked or when we were doing it. But year by year, as more ideas worked out and more smart people came into our orbit, we learned how to talk about what we were doing, and began to see the larger tradition we were part of.

In 2011, we decided to hold our first MAD Symposium (*mad* is the Danish word for "food"), a gathering of a few hundred

10

people with a vested interest in seeing the food world get better: people from the restaurant trade, along with scientists, farmers, philosophers, and artists. We chose the theme "Planting Thoughts," and we began thinking of potential speakers who could bring diverse thoughts about the plant kingdom.

I'll be honest with you: David Chang immediately came to mind because of kimchi. He may not remember serving it, but I remember having an oyster topped with kimchi water at Momofuku Ssäm Bar and finding it absolutely incredible. He and his team were working a parallel track to our own, learning their way around fermentation and developing new products using age-old techniques. I asked him to come speak at MAD about fermentation. While onstage, he introduced the culinary community to the concept of *microbial terroir*.

Chang was referring to the largely unseen world of mold, yeast, and bacteria responsible for fermentation. They are omnipresent, transcending countless cultures and culinary traditions. What Chang was saying was that the microbes indigenous to any given region will always have their say in the flavor of the final product, in the same way that soil, weather, and geography affect wine.

At the time, people were talking about Noma as the restaurant responsible for defining modern Nordic cuisine. From our perspective, we felt saddled with a tremendous responsibility. How could we claim to be cooking Nordic food if we used techniques from abroad? The notion of microbial terroir helped change everything for us. Fermentation knows no borders. It's as much a part of the cooking tradition in Denmark as it is in Italy or Japan or China. Without fermentation, there is no kimchi, no fluffy sourdough bread, no Parmigiano, no wine or beer or spirits, no pickles, no soy sauce. There is no pickled herring or rye bread. Without fermentation, there is no Noma.

People have always associated our restaurant closely with wild food and foraging, but the truth is that the defining pillar of Noma is fermentation. That's not to say that our food is especially funky or salty or sour or any of the other tastes that

12

people associate with fermentation. It's not like that. Try to picture French cooking without wine, or Japanese cuisine without shoyu and miso. It's the same for us when we think about our own food. My hope is that even if you've never eaten at Noma, by the time you've finished reading this book and made a few of the recipes, you'll know what I mean. Fermentation isn't responsible for one specific taste at Noma—it's responsible for improving everything.

It was with that in mind that in 2014 I asked Lars Williams and Arielle Johnson to build a space dedicated to exploring fermentation. Lars was one of our longest-tenured chefs, and Arielle became our resident scientist in 2013 while finishing her PhD in flavor chemistry. The two of them were responsible for taking our efforts to the next level, turning fermentation into a pursuit of its own at Noma—almost completely separate from the day-to-day activities of running the restaurant.

I was inspired by what the chefs at El Bulli had done in separating the actual creative part of their work from the service kitchen. Research and development weren't just activities to be done in between preparing mise en place and cooking for service. There was a team dedicated to them. That changed the game for creative cooking, and that's what we wanted to do for fermentation at Noma.

During the summer break at Noma, Lars and Arielle began planning what their ideal fermentation lab would include (within reason, of course). Up until then, we'd been fermenting wherever we could—on the boat, in the rafters of adjacent buildings, in old refrigerators, under desks.

They came back after a week or two and said the cheapest and most efficient way to do it would be in shipping containers. Things came together quickly. One day, three huge containers came in by forklift and crane. The team insulated the interiors and put up walls and doors. Lars went to Ikea, bought the second-cheapest kitchen, and merged it with equipment we'd amassed over the past decade. We started planning in June or July, and by August we had our fermentation lab.

13

I mention all this because I don't want to over-romanticize fermentation. It can be a pain in the ass to get everything up and running. It's work, but it's incredibly gratifying work. It's actually an amazing feeling to wait for something to ferment. It runs totally contrary to the spirit of the modern day.

And once you have your first ferments, it makes cooking so much easier. I really mean that. Some of these ferments are like a perfect cross between MSG, lemon juice, sugar, and salt. They can be drizzled onto cooked greens, added to soups, or blended into sauces. You can smear lacto-fermented plums onto cooked meats, or use the juice to dress raw seafood. And homemade ferments, packed into glass jars, make for unique and impressive gifts. Once you integrate these ingredients into your cooking, your eating life is going to be irreversibly better.

—

David Zilber started working with us the same year we built the fermentation lab. He came to us as a cook from Canada, and started in the restaurant as a chef de partie. When Lars and Arielle were leaving Noma in 2016, I was a bit distraught that we'd have to find someone to take over their work in the lab. But our head chef at the time, Dan Giusti, said we wouldn't have to look far. We installed David as the head of the fermentation lab,

and it's been a perfect fit. He has an incredibly quick mind and an insatiable curiosity. He understands the science underlying fermentation, and brings a line cook's work ethic to its practice. If you ask him something he doesn't know the answer to, rest assured he'll be completely educated about it the next time you talk to him. He's like a machine designed specifically to write this book with me.

And it's important to me that this book exist. It's important that we document the good work that people have done here. But I'm most excited by the prospect of people taking that work and applying it outside the restaurant. We've written books before, but none where the main goal was to translate what we do in the restaurant to a home kitchen. It's exhilarating to think that people all around the world will be able to get a sense for how we cook at Noma.

That's the only possible next step for what we've been working on this past decade. Restaurants have influenced what's sold on grocery store shelves. They've invigorated tourism in regions like ours, where people would never have thought to come eat before. The next phase is more education and more cooking— people connecting what we do at top-level restaurants with their everyday lives. That's how we can create a completely new culture of eating.

15

16

The broth is made by braising sea snails in an oil made from dried koji, then combining the cooking liquid with seaweed stock and more oil. It's served in the shell, garnished with pickled herbs.

At this point, the rate of discovery in the fermentation lab has slowed. We continue to adapt techniques to different ingredients, and some ferments remain less explored than others, but we're not stumbling into eye-opening new products at the same pace. When you've made garums (ancient fish sauces you'll learn all about later) from every type of seafood in Scandinavia, and they've all been good, it becomes difficult to identify the nuances. By putting this knowledge out there, we're hoping that not only will readers experience the same joy of discovery as we have, but that we'll get something out of it, too. We hope this will spur on the field. Perhaps one of you will take what you've learned here and come up with something completely new. If we're lucky, that will come back to Noma and bolster us.

I believe in fermentation wholeheartedly, not only as a way to unlock flavors, but also as a way of making food that feels good to eat. People argue over the correlation between fermented foods and an active gut health. But there's no denying that I personally feel better eating a diet full of fermented products. When I was growing up, eating at the best restaurants meant feeling sick and full for days, because supposedly everything tasty had to be fatty, salty, and sugary. I dream about the restaurants of the future, where you go not just for an injection of new flavors and experiences, but for something that's really positive for your mind and body.

I hope this book can be a launching pad for home cooks and restaurant cooks alike. When we think of our ideal readers, David and I talk equally about the parent who's passionate about cooking for his or her family and doesn't mind a weekend project, as well as the professional cook or sous-chef who can read between the lines and pull out novel ideas.

Studying the science and history of fermentation, learning to do it ourselves, adapting it to local ingredients, and cooking with the results changed everything at Noma. Once you've done the same and have these incredible products at your disposal—whether it's lacto-fermented fruit, barley miso, koji, or a roasted chicken wing garum—cooking gets easier while your food becomes more complex, nuanced, and delicious.

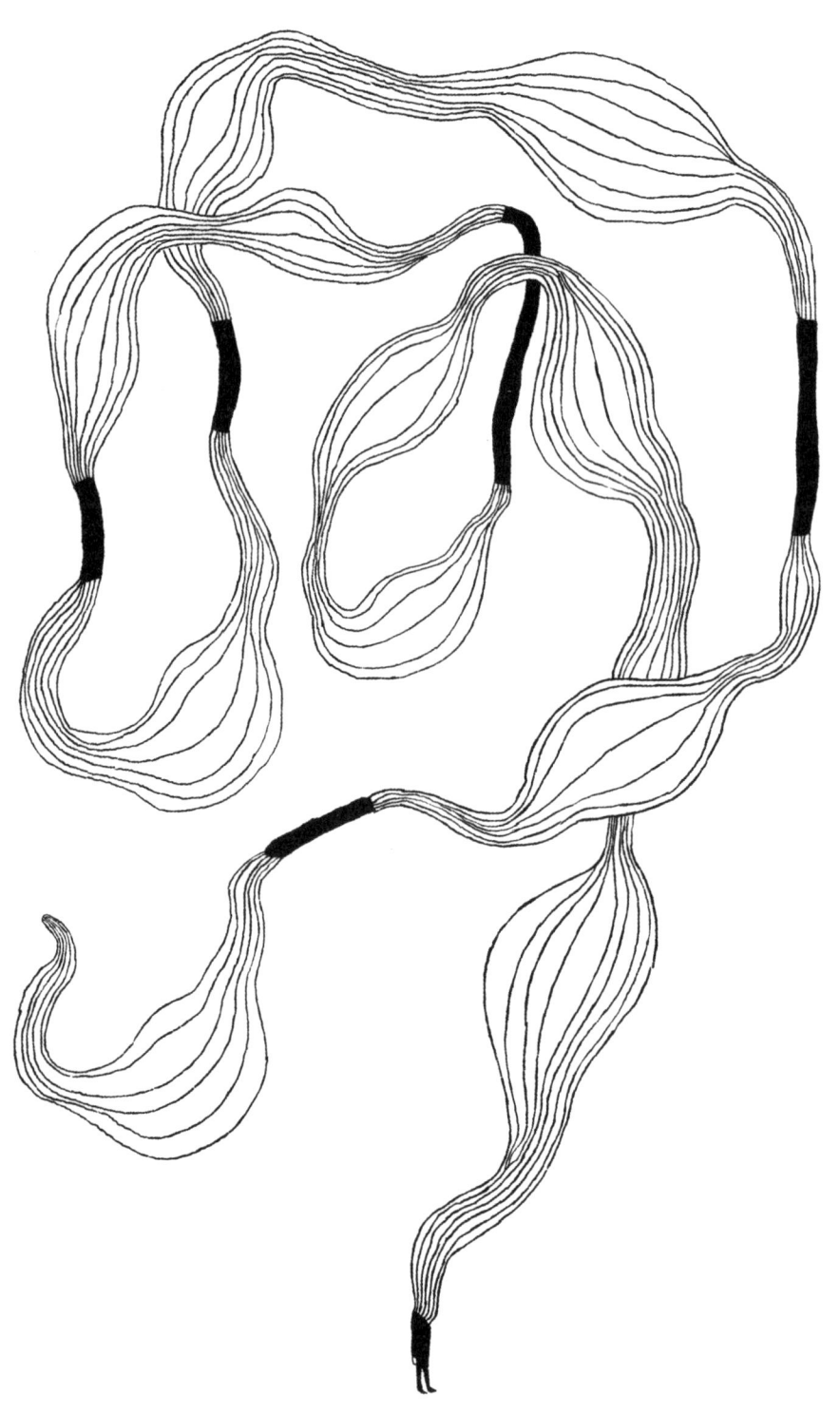

18

About This Book

There are thousands of products of fermentation, from beer and wine to cheese to kimchi to soy sauce. They're all dramatically different creations, of course, but they're unified by the same basic process. Microbes—bacteria, molds, yeasts, or a combination thereof—break down or convert the molecules in food, producing new flavors as a result. Take lacto-fermented pickles, for instance, where bacteria consume sugar and generate lactic acid, souring the vegetables and the brine in which they sit, simultaneously preserving them and rendering them more delicious. Cascades of secondary reactions contribute layers of flavors and aromas that didn't exist in the original, unfermented product. The best ferments still retain much of their original character, whether that's a touch of residual sweetness in a carrot vinegar or the floral perfume of wild roses in a rose kombucha, while simultaneously being transformed into something entirely new.

This book is a comprehensive tour of the ferments we employ at Noma, but it is by no means an encyclopedic guide to all the various directions you can take fermentation. It is limited to seven types of fermentation that have become indispensable to our kitchen: lactic acid fermentation, kombucha, vinegar, koji, miso, shoyu, and garum. It also covers "black" fruits and vegetables, which aren't technically products of fermentation but share a lot in common as far as how they're made and used in our kitchen.

Notably absent from this book are investigations of alcoholic fermentation and charcuterie, dairy, and bread. (Bread could take up—and deserves—its own separate discussion.) While we dabble with the fermentation of sugar into alcohol, it is almost always en route to something else, like vinegar. We've always worked closely with incredible winemakers and brewers and cannot pretend to be masters of their domain. Charcuterie is something that has not yet played a large role in our menus, though over the coming years we intend to dive deeper into fermenting meats as we celebrate the game season each fall. While we do make cheese at the restaurant, it's often served fresh and unfermented (though we're no strangers to yogurt and crème fraîche). Whenever we have cooked with artisanal aged cheeses, we've left their production in the hands of Scandinavia's amazing dairy farmers.

Each chapter tackles one ferment, providing some historical context and an exploration of the scientific mechanisms at work. Many of the ideas and microbial players behind different ferments are interconnected, so you'll see some concepts revisited and developed over the course of the book. For example, in order to make shoyu, miso, and garum, you'll first need to understand how to make koji, a delicious mold grown on cooked grains and harnessed for its powerful enzymes. That being said, you should feel free to dive in wherever your interests lead you. You'll still get a thorough understanding of each ferment without reading the rest of the book.

Included with each chapter is an in-depth base recipe, where we put ideas to work and walk you through the steps of making a representative example of each style of ferment. In most cases, there's no single "right" way, so the recipes are written with multiple methods and possible pitfalls in mind. We go into quite a bit of detail—more than you may need in some instances—but we want you to feel as comfortable making these ferments as one of our own chefs would be if tasked with making one for the first time. Even though it may require a little patience and commitment, you can and absolutely should produce your own shoyus and misos and garums. Once you taste the rewards of your effort, it's hard to imagine cooking without them. Plus, it all gets easier the second time around.

After you've read the in-depth base recipe for a ferment, you may feel ready to apply the same process to other ingredients, but to give you some inspiration, each chapter also contains several variations, which may illuminate other facets of the same technique. In some cases, these variations diverge in method from the base recipe, but rest assured, we'll detail these changes and explain why we're making them.

Finally, following each recipe, you'll see a few practical applications for the ferment in your day-to-day cooking—many

Roasted Bone Marrow, Noma, 2015

The bone marrow is marinated in beef garum and
elderberry vinegar, then roasted over coals. It's served
with cabbage leaves dressed with an emulsion of
caramelized beef garum pulp and a sauce of white
currant juice seasoned with lacto cep water.

21

Chilled Oysters and Salted Green
Gooseberries, Noma, 2010

A lightly poached Danish oyster is
dressed with slivers of lacto-fermented
green gooseberries and their juice.

of which are inspired by preparations we make at Noma. Think of them as things that a cook from Noma would make for dinner at home using the ferments in the book. We've written these short recipes in a more informal manner, taking a cue from the naturalist Euell Gibbons, who wrote beautifully about foraging—another preoccupation of ours. In his book *Stalking the Wild Asparagus*, Gibbons details how to identify and harvest wild plants, and then provides recipes in a fluid, conversational format—suggesting rather than prescribing what to do with the incredible ingredients you can find outdoors. It's the same approach we're trying to take here. We don't go into step-by-step detail when it comes to how you can employ the ferments in this book, because the specifics aren't nearly as important as the possibilities. Even if you don't feel up to making your own ferments, you'll still find all manner of new uses for store-bought versions.

This is a book meant to bring some clarity to a hazy realm of cooking, full of confusing and unfamiliar terminology. We've spent the past decade investigating and unraveling fermentation for ourselves, and we'll try to share what we've learned with you. But more important, we want you to come away from this book with the same feeling of exhilaration and wonderment that we have whenever we make and use one of the miraculous products of fermentation.

23

1.

Primer

—

What *Is* Fermentation? 26

What Makes Fermentation
Delicious? 27

Setting the Table for Microbes 29

Wild Fermentation 33

Backslopping 33

Cleanliness, Pathogens,
and Safety 36

Potential of Hydrogen (pH) 40

Salt and Baker's Percentages 41

Building a Fermentation
Chamber 42

Thinking Outside the Kraut 50

Substituting Store-Bought
Ferments 51

Weights and Measures 52

What *Is* Fermentation?

Before we dive into the practical ins and outs of fermentation, let's first clearly define what it is.

At the most basic level, fermentation is the transformation of food by microorganisms—whether bacteria, yeasts, or mold. To be slightly more specific, it is the transformation of food through enzymes produced by those microorganisms. And finally, in the strictest scientific definition, fermentation is the process by which a microorganism converts sugar into another substance in the absence of oxygen.

The word *fermentation* comes from the Latin word *fervere*, meaning "to boil." The ancient Romans, upon seeing vats of grapes spontaneously bubble and transform into wine, described the process using the closest analogue they could think of. And while those bubbling vats of grapes had nothing to do with boiling, they *were* true ferments in the scientific sense, as yeast-produced enzymes transformed the sugars in the grapes into alcohol.

However, not all the processes we consider to be fermentation fit neatly into tidy definitions of it. For instance, while koji is faithful to the definition, Noma's garums are not. In koji, the mold *Aspergillus oryzae* penetrates grains of rice or barley and produces enzymes that convert the grain's starches into

You taste as much with your brain as you do with your tongue.

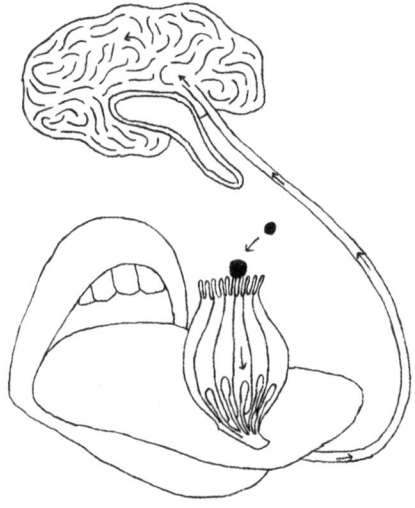

simple sugars and other metabolites. This is what's known as a *primary* fermentation process. The garums in this book, on the other hand, are the product of a *secondary* fermentation process. To produce garum, we mix koji with animal proteins in order to take advantage of the enzymes produced during the primary fermentation process.

We don't differentiate between primary and secondary fermentation processes in this book, but you may find it helpful to have these definitions under your belt as you find your way with fermentation.

What Makes Fermentation Delicious?

Taste is a function of the human body, and to understand what tastes good to us, we have to understand its role in our evolutionary history. All our senses serve to aid in our survival. Our senses of taste and smell have been shaped over hundreds of millions of years to incentivize us to eat foods that are beneficial to our bodies. Our tongues and olfactory system are unbelievably complicated organs that take in chemical cues from the world around us and transmit that information to our brains. Taste lets us know that a ripe piece of fruit is sweet and thus full of calorie-rich sugar, or that a plant's stalk is bitter and potentially poisonous. We are born with aversions to certain flavors (a sense that becomes reinforced by experience), leading us to gag at the stench of rotting flesh decaying at the hands of pathogenic bacteria, while we register the scent of meat roasting over fire as mouthwateringly delicious, because it indicates to our brains that we're about to eat something rich in proteins.

There are numerous biological processes at work in any given fermentation, but the ones that matter most to us from a taste perspective are those that break down large chains of molecules into their constituent parts. The starches in foods like rice, barley, peas, and bread are actually long chains of linked molecules of glucose, a simple sugar. Proteins, which can be found in large supply in soybeans and meat, are constructed in a similar fashion from lengthy, winding chains of amino acids—small organic molecules essential to all aspects of life on earth. One of those amino acids, glutamic acid,

27

registers on our taste receptors as umami—the elusive, crave-able quality that connects foods like mushrooms, tomatoes, cheese, meat, and soy sauce.

So what makes fermentation so good? On their own, starch and protein molecules are too large for our bodies to register as sweet or umami-rich. However, once broken down into simple sugars and free amino acids through fermentation, foods become more obviously delicious. Koji made from rice has an intense sweetness that plain cooked rice doesn't. Raw beef left to ferment into garum has a savoriness that speaks to us on a primitive level.

Simply put, the microbes responsible for fermentation transform more complicated foodstuffs into the raw material your body needs, rendering them more easily digestible, nutritious, and delicious. Our affection for the tastes those microbes produce has allowed them to evolve and stay in our company. Humans have been fermenting for so long that many of the microscopic agents responsible can be considered domesticated, just like household cats or dogs. But while pets can stare longingly at you if they're hungry or cold, microbes are a bit trickier to read. It's a mutually beneficial relationship, but one that needs a bit of work to keep everyone happy. That's the job of the fermenter.

Proteins are made of tangled chains of amino acids, life's building blocks.

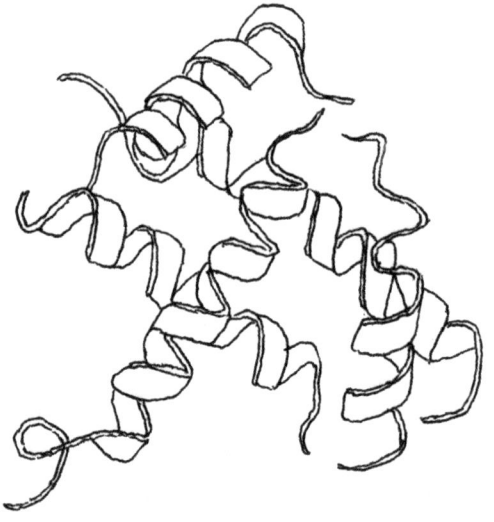

The number of species of microbes
on earth is greater than that
of all plants and animals combined.

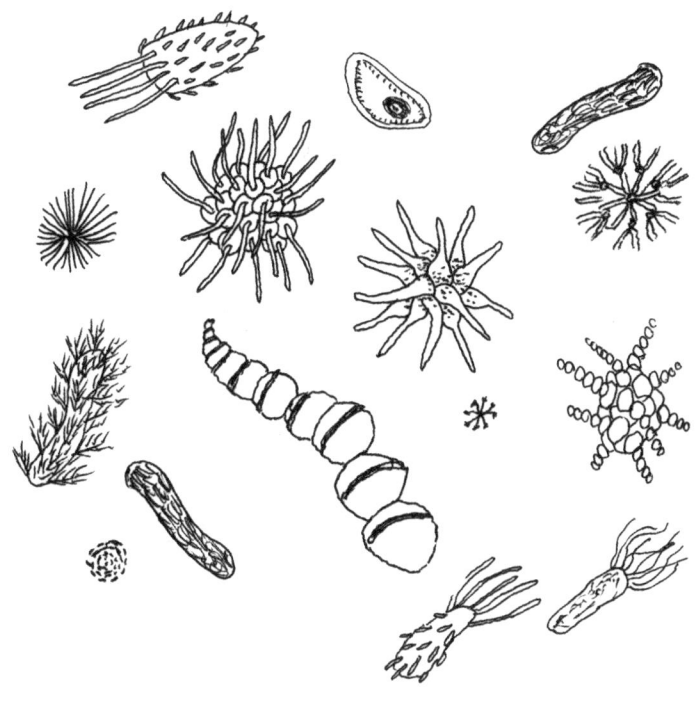

Setting the Table for Microbes

There's a thin line between rot and fermentation, and that line might best be understood as an actual line, like the kind you'd find outside a nightclub. Rot is a club where everyone gets in: bacteria and fungi, safe or unsafe, flavor enhancing or destructive. When you ferment something, you're taking on the role of a bouncer, keeping out unwanted microbes and letting in the ones that are going to make the party pop.

You have several tools at your disposal in trying to encourage certain microbes or deter others. Some organisms are more tolerant of acidity than others. Likewise with oxygen, heat, and salinity. If you're familiar with what your preferred microbe needs to function, you can wield these factors to your benefit. Each chapter in this book will go into great detail about the conditions you need to create successful fermentation, but for starters, here's an overview of the players that will be working for us.

29

Bacteria

Among the earliest forms of life, bacteria are single-celled organisms that are present in uncountable quantities in nearly every corner of the globe. Only a fraction are known to science. There are malignant bacteria that can produce toxins capable of killing much larger organisms. At the same time, there are billions of beneficial bacteria living on and inside of us. At the end of the day, the majority of them are harmless to us.

Lactic acid bacteria (LAB)

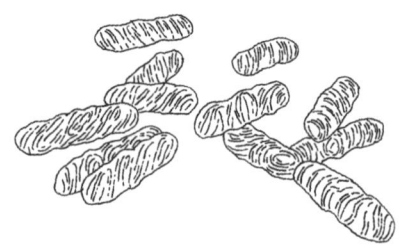

LAB are rod- and sphere-shaped bacteria that are present in abundance on the skins of fruits, vegetables, and humans. We use them for their ability to convert sugar into lactic acid, giving pickles, kimchi, and other lacto-fermented products their characteristic sourness. Because they produce lactic acid, they are able to tolerate low-pH environments. They are also halo-tolerant (salt-tolerant) and anaerobic, meaning they thrive in the absence of oxygen.

Acetic acid bacteria (AAB)

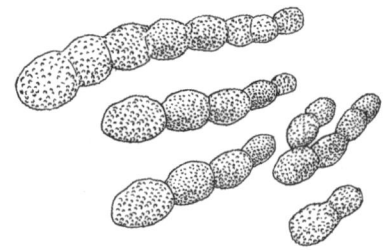

Like LAB, AAB are readily abundant rod-shaped bacteria, ever present on the surface of many foods. They generate the sharp sourness of vinegar and kombucha by converting alcohol to acetic acid. We often use them in conjunction with yeasts that first convert sugars into alcohol. They can tolerate the acidic environments they create, and require oxygen to create acetic acid, thus classifying them as aerobic bacteria.

Fungi

Fungi encompass a huge swath of life on earth, from single-celled yeasts to molds to gigantic puffball mushrooms. Multicellular, filamentous fungi like mushrooms and molds grow by gathering nutrients through tendril-like hyphae that together form a web-like system known as a mycelium, similar to the roots of a plant. They secrete enzymes through their mycelium, effectively digesting the food in their surroundings, then absorbing the nutrients from their environment.

Saccharomyces cerevisiae

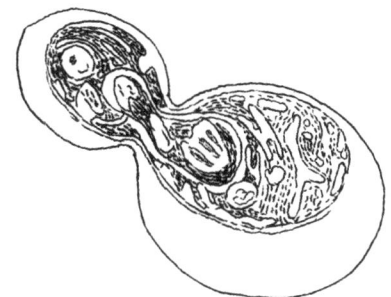

An extremely handy species of yeast, *Saccharomyces cerevisiae* is responsible for three of humanity's most important culinary pillars: bread, beer, and wine. Bountiful in the natural world, as demonstrated by producers of spontaneously fermented bread and wine, *S. cerevisiae* makes a living converting sugars into alcohol. It breaks down glucose to harness the chemical energy needed for its life processes, while producing carbon dioxide and ethanol as by-products. Different strains or subspecies are harnessed for their particular qualities, which can lead to wide variations in flavor. For instance, the strain of *S. cerevisiae* that is used in bread baking isn't desirable for producing beer or wine. Yeast can survive and multiply in the presence of oxygen, but alcohol fermentation takes place anaerobically. *Saccharomyces* dies at temperatures in excess of 60°C/140°F.

Brettanomyces

A genus of long, cylindrical yeast, *Brettanomyces* is used in the production of beers with sour qualities because of its ability to produce acetic acid as a metabolite. *Brettanomyces* also occurs naturally on the skins of fruits, and can be purchased readily as "saison yeast." It can survive in oxygen, but produces ethanol anaerobically. Like other yeasts, it cannot survive temperatures above 60°C/140°F.

31

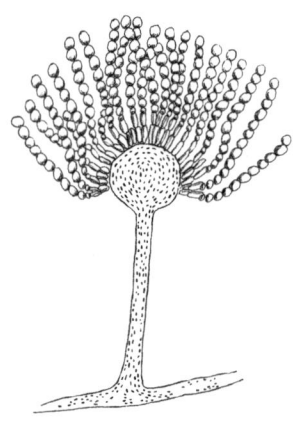

Aspergillus oryzae

Perhaps the most important microbe in this book, *A. oryzae* (pronounced oh-RAI-zee) is the sporulating mold also known as koji. It's been bred for hundreds of years to grow extremely quickly in hot and humid environments when given access to the plentiful starches in products like cooked rice or barley. (Generally speaking, 30°C/86°F and 70% to 80% humidity are ideal for *Aspergillus*; temperatures above 42°C/108°F will kill it.) Koji secretes the enzymes protease, amylase, and a small amount of lipase, which break down proteins, starches, and fats, respectively. We harness these enzymes in the production of our misos, shoyus, and garums.

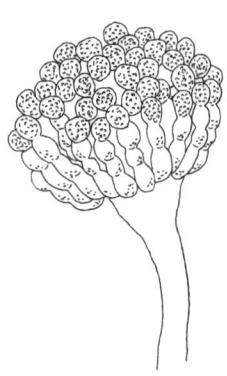

Aspergillus luchuensis

A relative of *Aspergillus oryzae*, *Aspergillus luchuensis* (pronounced loo-CHOO-en-sis) metabolizes starches and proteins and produces citric acid as a by-product. It's traditionally used to brew the bases of Asian spirits like Korean *soju* and Japanese *awamori*, as the distillation of the alcohol leaves the citric acid behind. Though it's a lesser-known species, it's extremely delicious.

Enzymes

Enzymes are not microbes—they aren't even alive—but rather biological catalysts that facilitate chemical transformations within organisms or organic matter. You can generally identify them by the suffix *-ase*, as in protease (an enzyme that breaks down proteins) and amylase (from the Latin word *amylum*, meaning "starch," which breaks down exactly that). They are a class of proteins built through evolution to serve specific but different functions. Exactly how they work is rather complicated, but you can think of the ones featured in this book as a cross between keys and scissors. They're keys in the sense that they are tailored to fit specific locks, acting on one organic molecule while leaving others alone; and they're scissors in that they can cut ribbons into shorter lengths. Generally speaking, enzymes work most efficiently in warm, fluid environments, but if heated too high, they can be "cooked" to a point where they no longer function.

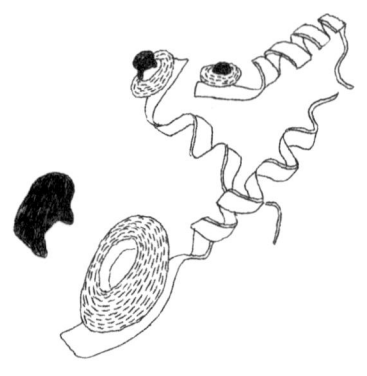

Beta-amylase is an enzyme capable of breaking down starches into their constituent sugar molecules.

Wild Fermentation

The ferments we undertake at Noma all depend to varying degrees on wild fermentation. That is to say, we create environments that are conducive to the growth of naturally occurring beneficial microbes, and detrimental to malevolent ones. With our lacto-ferments, for instance, we depend entirely on a wide set of lactic acid bacteria in the environment—on the fruit or vegetables we're fermenting, on our hands, floating in the air—to turn sugar into lactic acid and other flavorful metabolites. By allowing nature to do its thing, we get layers of nuance and complexity in our ferments that wouldn't be possible if we dictated exactly which microbes were allowed to work. Wild fermentation is a non-inoculated and often very diverse fermentation. Simply put, it's how fermentation was first performed, and it's still tried and true.

For our kombuchas, vinegars, and koji, we do introduce bacteria, yeast, or fungus into the equation in order to get the results we're looking for, but we still allow and encourage wild fermentation. The same goes for especially large batches of lacto-fermented products. For instance, when we're fermenting hundreds of kilos of asparagus at a time, we add powdered lactic acid bacteria (LAB) to the brine. If for some reason the naturally occurring LAB had trouble getting started, we'd be exposed to the risk of some other malignant microbe taking hold. A boost in the LAB population is a nice bit of insurance against losing all that product when you're working on a large scale.

Backslopping

Backslopping is a vital technique in prepping microbial environments for fermentation and will come up numerous times in this book, especially in the production of kombucha and vinegar. The idea is basically to give the substance you intend to ferment a boost of beneficial microbes by adding a dose from a previous batch of that same ferment.

By pouring a healthy amount of, say, perry vinegar into a jar of fresh perry, we both lower the pH of the solution and add a healthy shot of acetic acid bacteria (AAB). Lowering the pH (acidifying) has the effect of slowing or stopping any unwanted microbes that aren't acid-tolerant from acting on the perry,

33

Backslopping gives a boost
from one generation of a ferment
to the next.

and ensures that there's a healthy population of AAB to ferment the perry into perry vinegar. Backslopping stacks the deck in favor of the microbes we want to succeed.

Of course, if this is your first time making one of the ferments in the book, you won't necessarily have a previous batch to use for backslop. In that case, you'll have to find a similar substitute. For our vinegars, we suggest unpasteurized apple cider vinegar as a replacement. For our kombuchas, you can use a similarly flavored unpasteurized kombucha or the liquid that your SCOBY (the "mother" culture of yeast and bacteria that produces kombucha; see page 111) comes packaged in. The downside is that you're going to dilute the pure flavor of the vinegar or kombucha you're making. That's fine, though, as it gives you a perfect reason to make the same vinegar or kombucha again—this time using a portion of your first batch as backslop.

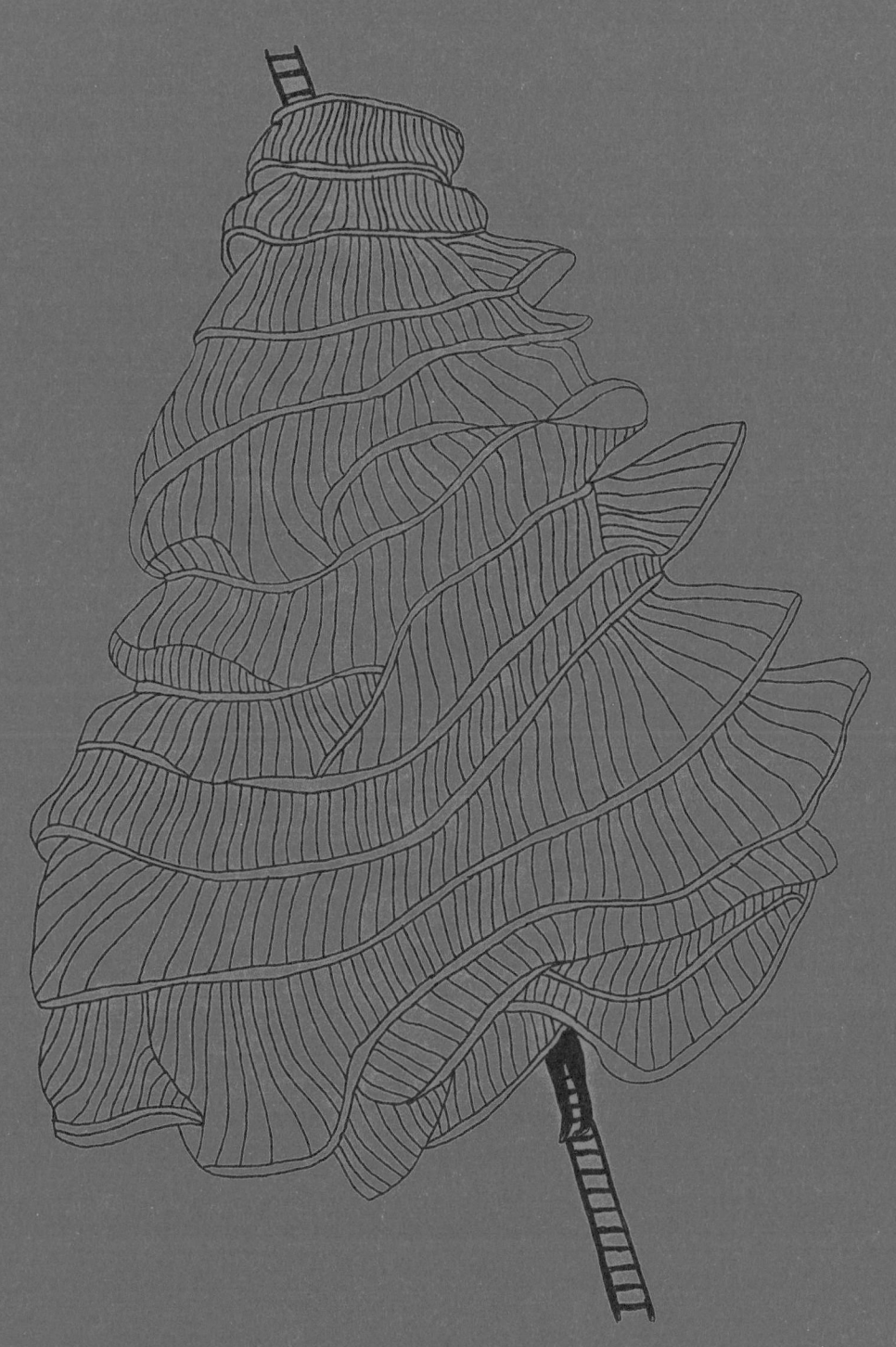

Cleanliness, Pathogens, and Safety

Cleanliness is something we take very seriously in the kitchen, out of both pride for our workplace and respect for our colleagues. However, a clean and sanitary workplace is doubly important in the fermentation lab, in order to prevent unwanted pathogens from invading a ferment and causing it to taste off or, worse, become dangerous to eat. At Noma, we always err on the side of caution. If something you've made smells *wrong*—not just funky like fish sauce, but nose-stingingly rotten—trust your nose. If you taste a small sample and it turns your stomach, remember that your body is designed to reject things that may be harmful to you. When in doubt, throw it out. If you're ever unsure of a fermented product, toss it. The weeks or months of your invested time are not worth risking your health.

Potentially harmful microbes are ever present in the environment. Bacteria can multiply speedily, with or without oxygen, at temperatures ranging from 4.5° to 50°C/40° to 122°F, especially in moist, nutrient-rich environments. Of course, that describes the exact circumstances in which many fermented goods are produced. Both the World Health Organization and the United States Department of Agriculture recommend cooking foods sensitive to pathogenic contamination above 70°C/158°F before consumption. Now, that's a fairly severe safeguard, and obviously not possible for many ferments. That being said, you should be cautious, but not worried. Fermentation is meant to be a rewarding and exhilarating practice, but remember that you're playing with live ammo.

Throughout this book, we do our best to provide clear instructions that will produce safe and delicious products if followed closely. Don't eyeball measurements or take shortcuts. When a recipe calls for a specific salt content (above 10 percent by weight) or pH (below 4.5), it's to ensure that you're fermenting safely. But of course, the first step in preventing unwanted microorganisms from taking hold in a ferment is to make sure your equipment and hands are clean before they come into contact with food. While this is less important in certain cases, it's critical in other instances. When making koji, for example, you'll need to be sure the incubation chamber is properly

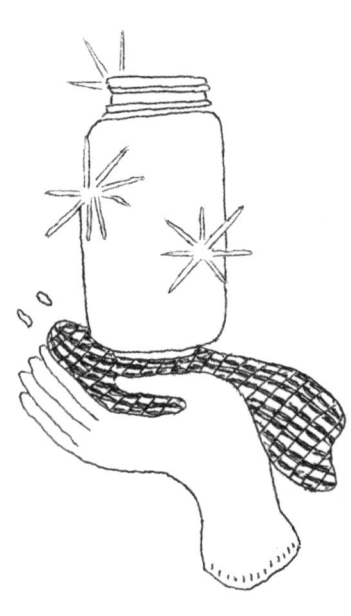

Cleanliness is next to godliness (and also crucial to a safe and successful ferment).

sanitized before introducing the inoculated grains. And when working with your hands, wear nitrile or latex gloves to prevent contamination (except in places where a little bacteria from your skin can help things along, as with lactic-acid fermentation).

Now, what do we mean by "clean"? There is a difference between the level of cleanliness you would expect to find in a university biology lab and that in a home or restaurant kitchen. Let's define some terms. *Cleaning* means that you've removed visible dirt from the surface of objects. Soap and water will clean a surface but do very little to reduce the surface's population of microorganisms, good or bad. *Sterilized* implies that you've eradicated *all* life-forms—viruses, bacteria, fungi—on your equipment and your work surfaces (and sometimes even in the product you're looking to ferment). This is a level of certainty required in hospitals and microbiology labs. You'll never need something as serious as an industrial-strength autoclave for a recipe in this book. What we're looking to do for the recipes here is *sanitize*. To sanitize a piece of equipment or work surface implies that you've removed *most* microbiological life. That will be sufficient for our purposes. Running your equipment through a hot cycle in a dishwasher or steaming or boiling it for a few minutes is more than enough to ensure that you're working clean and sanitarily. If your equipment is heatproof, dry-heat sterilization is another option. Ceramic, glass, and metal containers and utensils can be baked in the oven for 2 hours at 160°C/320°F to ensure that they're free of contaminants.

For equipment or work surfaces that you can't pop into the dishwasher, there are common sanitizers intended for food production and fermentation like StarSan (available at many home-brew shops), distilled white vinegar (a sanitizing agent favored by grandmas the world over), and even household bleach diluted with water to 20 milliliters per liter (as long as you rinse with fresh water afterward). At Noma, for large items like crocks and buckets, we disinfect using ethanol diluted with filtered water to 60 percent alcohol by volume (ABV)—40 milliliters water for every 60 milliliters ethanol.

37

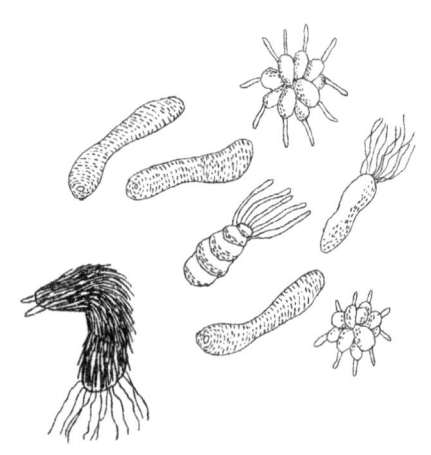

While many microbes are beneficial and the majority are harmless, there are still a few bad microbes that can cause illness.

(We dilute it because if the percentage of ethanol is too high, it can actually coagulate the proteins that make up the cell walls of many microbes and prevent them from dying.) We put the solution in a spray bottle and spray whatever needs to be sanitized, let it sit for 10 to 15 minutes, then wipe it off with a paper towel.

Finally, while a great deal of time is spent in this book introducing the amazing microorganisms responsible for fermentation, it's equally important to acquaint ourselves with the microbes that can make things go sideways. With a thorough grasp of pathogenic bacteria and molds, and what conditions they can tolerate, you'll be better equipped to keep them out of your products.

Clostridium botulinum

C. botulinum is the sporulating bacteria responsible for botulism. It is an anaerobic bacteria that thrives in nutrient-rich, warm environments. Its spores are commonly found dormant in soil and water, waiting for favorable conditions to propagate and release potent neurotoxins. Ingesting just a microgram of botulism toxin is enough to cause serious illness. You cannot taste or smell botulism toxin, and thus the only way to guarantee safety is through careful attention to best practices.

Though cases of botulism poisoning are rare, it's usually found in improperly refrigerated animal products or improperly canned vegetable products (where canning temperatures were not hot enough and/or the canning liquid was not sufficiently acidic). Given that the spores of the bacteria are often found in the soil, special attention should be paid when fermenting roots, bulbs, and tubers. When making black garlic, for example, you're keeping a root vegetable in an anaerobic environment at a warm temperature. However, *C. botulinum* cannot survive at a sustained temperature of 60°C/140°F. Your responsibility is to ensure that your heating chamber doesn't dip below that threshold.

C. botulinum also has great difficulty growing in fluid mediums with a water activity below 0.97 (achieved by salt concentrations of 5 percent or higher) and in acidic environments with a pH below 4.6. Many ferments in this book begin with salt concentrations lower than 5 percent and a pH above 4.6. However, the combined effect of moderate salt content and a gradually decreasing pH is usually enough to safeguard against malevolent bacteria. For instance, a vegetable brined at 2 percent salt will have a high enough salt content to inhibit *C. botulinum* while beneficial lactic acid bacteria lower the pH. If a ferment reaches a pH below 5 within the first two days and ends up below 4.6 by the time of completion, it is generally recognized as safe.

Escherichia coli

Many strains of *E. coli* are actually harmless and part of a normal gut flora, but some varieties can cause severe food poisoning. These bacteria are usually transmitted through poor hygiene or contaminated meat products. Cross-contamination of work surfaces and utensils is one of the more common causes of *E. coli*–related illness. Proper and thorough washing of vegetables in cold water will greatly reduce populations of the pathogen, should they be present. For products like beef garum, salt concentrations of 10 percent or higher will kill off the microbes. On top of that, the high temperatures at which garum ferments offer an added layer of protection.

Salmonella

Salmonella is a genus of rod-shaped bacteria often found in raw poultry products and unpasteurized milk and on unwashed fruits and vegetables. Doing everything you can to avoid cross-contamination from raw poultry is paramount in avoiding *Salmonella* food poisoning. For example, if you're cooking chicken wings for chicken wing garum, be sure to clean and sanitize any utensils before putting them back into action with the final, prepared ingredients. Like *E. coli*, *Salmonella* has a minimum water activity level of 0.95, meaning that salt levels above 10 percent will kill it off.

Pathogenic molds

There are thousands of wild and invasive molds that would jump at the opportunity to eat your fermentation project before

39

you get the chance. Many microscopic mold spores are airborne, while others travel in water or on the backs of insects. Not all of them will necessarily be harmful, but if you didn't put the mold there yourself, it's best not to take the chance.

There are many instances in this book where we are trying to create the ideal environment for beneficial mold growth, so the best preventative measures you can take against pathogenic molds are cleaning and sanitizing. By eliminating any unwanted guests at the outset, you ensure that they won't spoil the party later. Another method is to overwhelm competing molds. With koji, we inoculate steamed barley heavily with *A. oryzae* spores in order to elbow out the competition. With ferments like garums and shoyus, the salt content retards mold growth. Frequent stirring and cleaning of the container walls will bring any spores on the surface out of contact with the air and drown them in a salty sea. For kombucha, keeping the surface of your SCOBY moist by basting it with liquid is often enough to keep it acidified and mold-free. Last, molds are easier to spot than other pathogens. When making something like miso, you can simply scrape away any mold that forms on the surface.

Potential of Hydrogen (pH)

Potential of hydrogen, or pH, is a hugely important measurement in chemistry, and a key factor to consider in fermentation. Simply put, it helps you measure acidity. The pH scale was first conceived in the Carlsberg Labs in Copenhagen near the turn of the twentieth century. It measures the difference in concentration in an aqueous solution between hydrogen ions (H^+) and hydroxide ions (OH^-), with every increase in numerical value from 0 to 14 indicating a tenfold change in ionic concentration.

In distilled water (pure H_2O), hydrogen and hydroxide ions sit in exact balance with each other. It has a pH of 7, right in the middle of the scale, and is neither alkaline nor acidic, but neutral. When hydroxide ions outnumber hydrogen ions, the substance is said to be basic or alkaline, and has a pH above 7. When hydrogen ions outnumber hydroxide ions, the substance is acidic, and has a pH below 7. The most acidic substances

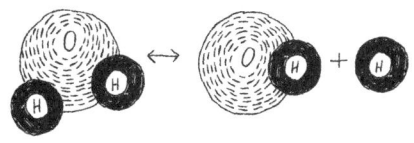

The ratio of hydroxide ions (negatively charged) to hydrogen ions (positively charged) in an aqueous solution determines its pH.

Salt and Baker's Percentages

you can find, like hydrochloric acid (a component of stomach acid) and sulphuric acid (found in car batteries), have a pH near 0. The most basic substances, like sodium hydroxide (found in lye or drain cleaner) have a pH close to 14.

At times in this book, we seek to control or change the pH of a ferment, which affects everything from microbes' ability to thrive and propagate to an enzyme's ability to function properly to the taste of the final ferment. Sometimes, we're seeking to lower the pH in a ferment—thus making it taste more sour—through the creation by microbes of lactic, acetic, or citric acid. We use alkaline solutions too, as in the case of our miso made from masa, where we boil corn in a calcium hydroxide solution to coax out floral and fruity notes from the kernels.

You can track pH using a few tools, including test strips or digital meters. More exacting fermenters may find these tools helpful, but taste is your best guide. Ultimately, what you find palatable should dictate what you think the "right" pH is.

Salt is one of the most important factors in a safe and successful fermentation. For starters, it has the remarkable ability to inhibit biological processes of both microbes and humans. (There's a reason why drinking salt water will kill you if you're stranded at sea.) Salt is an ionic compound of sodium and chloride, which breaks apart into a sea of ions when it dissolves in water. Nature abhors imbalance, so anywhere they can, water and the salt ions dissolved in it will try to spread out into an even distribution. Put a piece of meat or a bacterial cell in a solution of salt, and water from inside will flow outward while salt ions flow inward, until eventually equilibrium is reached. It's how brining works, and it's also the mechanism by which pathogens like *Salmonella* can be killed with salt. Salt draws water out of the bacteria's cells until they shrivel up and die. (For a more detailed explanation of this, see "Salt/Water," page 367.) Knowing the salt tolerance of different microbes can make a world of difference in a ferment.

For that reason, we stress precise salt measurements, usually expressed in percentage by weight. Note that in the fermen-

41

tation lab at Noma, we use baker's percentages—when we tell you to add 2% salt to a kilogram of plums, we mean 2% of the weight of the plums (which comes out to 20 grams), not the total weight of the plums *and* the salt (which would be 20.4 grams). The difference is not always very significant, but using baker's percentages streamlines the math.

Last, the type of salt makes a difference. We call for non-iodized salt, because iodine is mildly antimicrobial. While using standard table salt won't stop a ferment cold, it can impede helpful microbes from gaining a strong foothold. Kosher salt will work well, and should be available in your local grocery store. Mineral-rich sea salts like fleur de sel are great, too, and can actually improve the texture of lacto-ferments.

Beginning with the koji chapter, you'll find that some recipes in this book require specific temperature and humidity conditions. There are myriad options for constructing a fermentation chamber, depending on how much product you intend to make, and how elaborate you want your rig to be. At Noma, we have rooms dedicated to fermentation, with accurate and precise temperature and humidity controls. During our pop-up restaurant in Sydney, we crafted a fermentation chamber out of a broom closet. You can use a decommissioned refrigerator, a speed rack with a vinyl cover, Styrofoam coolers, or wooden boxes. The two basic criteria for a good container are insulation and water resistance. The chapter on koji (page 211) explains what factors you need to control and why they matter.

While you're getting your feet wet in the world of fermentation, an appliance such as a rice cooker or slow cooker will suffice for some processes in this book. (Note that you'll need a model without an auto-off function, as some recipes call for incubation times that last for weeks.) But once you're hooked on fermenting, building a larger, more accurate chamber is a game changer.

Here we've outlined two paths, designed for small-scale projects, built using components that are available online or at a hardware or restaurant supply store. It can all be done for less than the cost of a stand mixer.

Building a Fermentation Chamber

Covered Speed Rack

For this fermentation chamber, you'll need:

- A speed rack: The bones of your chamber. Speed racks are used in restaurants to hold trays of ingredients or food coming out of the oven. They're made of lightweight but sturdy aluminum and are equipped with rails onto which you slide sheet pans or gastro/hotel pans. They come in varying heights, ranging from 1 to 1.75 meters. Look for one that comes with a heavy plastic or vinyl cover with zippers running up two sides. The cover will retain heat and humidity, and the zippers allow easy access to the interior. You'll also need a few sheet pans that are the correct size for the rack; the style and quantity will depend on which ferments you choose to make.

- A small space heater: The kind you might use to keep your feet warm underneath your desk. If the heater is equipped with a fan, all the better; if not, buy a small simple fan.

- A temperature controller, such as a PID (proportional-integral-derivative) or thermostat: This will adjust the temperature of the chamber as it varies according to external influences. You want a prewired version that you can plug a heater directly into. It's a specialized bit of gear, but it's not complicated or expensive. It will include a probe that you set either in the chamber to measure interior temperature, or into the ferment itself, such as when you're making koji.

- A small humidifier (only when making koji): The type you'd put in a child's room to help with a stuffy nose. Plus, a simple hygrometer to gauge humidity; it will look a bit like an oven thermometer. Or use a humidistat, which functions much like a thermostat. While slightly more expensive, it will simplify things by regulating the humidity in the chamber for you.

43

Building a Fermentation Chamber with a Covered Speed Rack

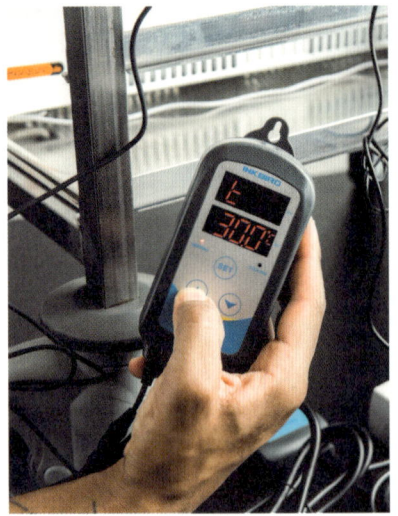

1. Assemble the speed rack and slide one or two sheet pans into the lower shelves. Space them to allow enough room for your heater, humidifier, and hygrometer or humidistat (and fan, if the heater doesn't have one built in) to sit without interfering with one another. Place the devices on a sheet pan and snake the cords out from the bottom of the rack.

2. You'll want to keep your temperature controller outside the chamber. Plug it in and set it to the correct temperature, following the manufacturer's instructions; for the ferments in this book, that will be either 30°C/86°F or 60°C/140°F. Run the temperature probe into the chamber. Plug the heater into the temperature controller.

3. Arrange the hygrometer or humidistat sensor so it won't be in the direct flow of the steam from the humidifier. Fill the humidifier with water, plug it in, and set it to medium. Note that we're clearly dealing with a lot of electrical cords, so use a properly rated power strip.

4. Pull the plastic cover over the speed rack and zip it up. Air will be able to enter the chamber from the bottom, which is what you need for most of the ferments. When fermenting at 60°C/140°F, you may want to add an extra layer of insulation underneath or on top of the plastic cover. A clean cotton or wool blanket will do the trick nicely.

5. Close the cover to bring your chamber up to the desired temperature and humidity. If you don't have a humidistat, you'll adjust humidity by checking the level on your hygrometer and then dialing the humidifier setting up or down. The temperature controller will take care of temperature for you.

6. Add your ferments. Keep an eye on the temperature controller to make sure it's turning the heater on and off when the temperature dips or rises. You may see a drift of a degree or two above or below your desired temperature; that's normal.

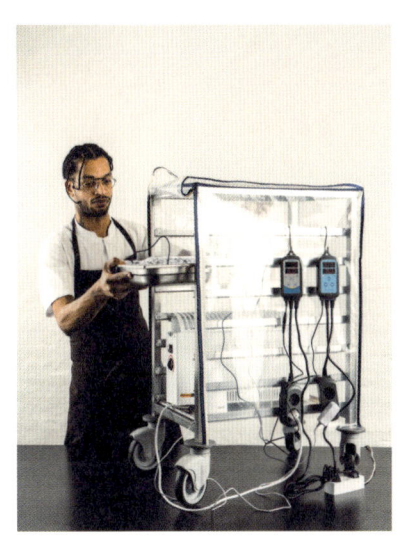

45

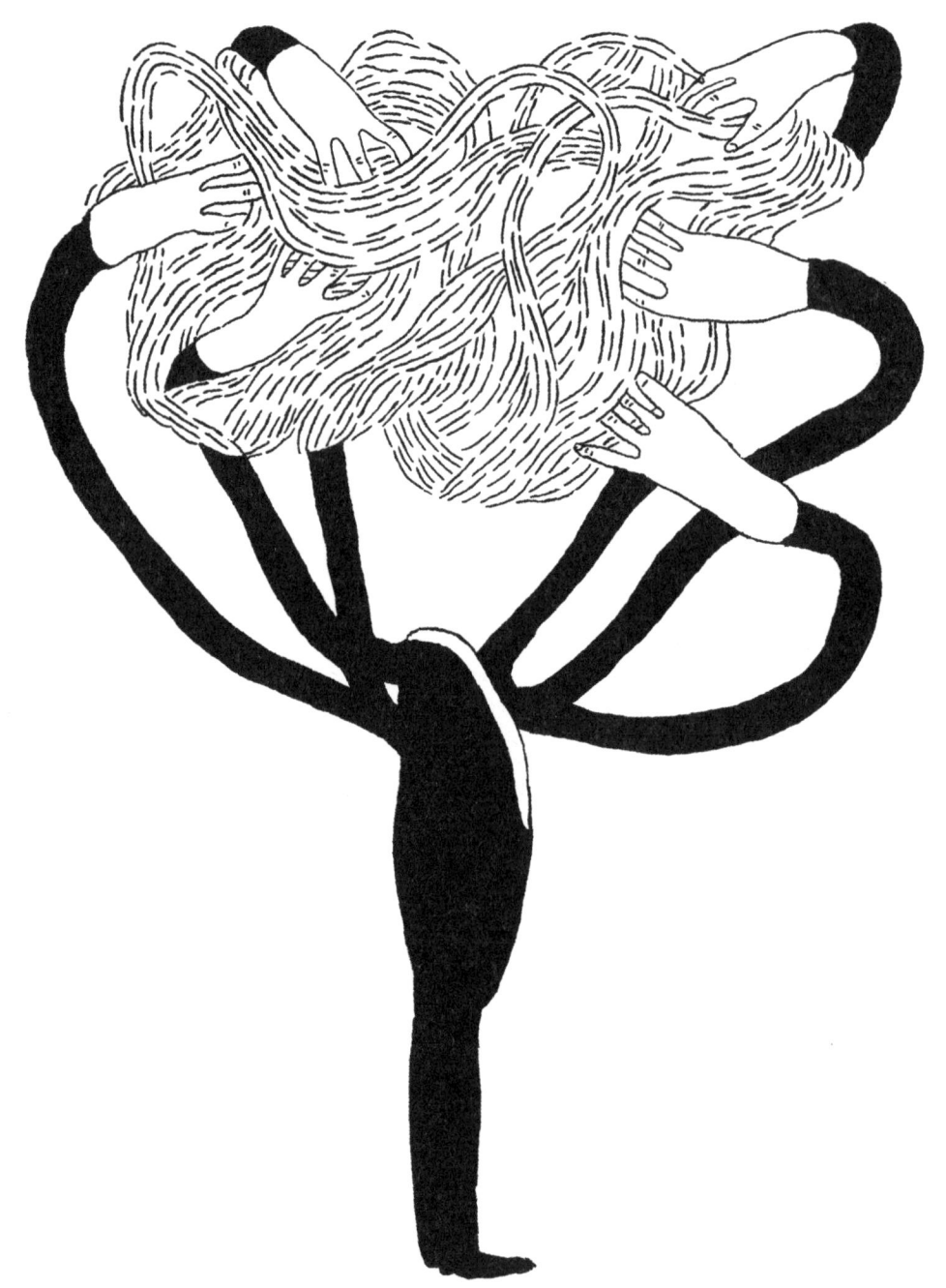

46

Styrofoam Cooler

For this fermentation chamber, you'll need:

- A Styrofoam cooler: Styrofoam is an excellent insulator, and Styrofoam coolers are fairly inexpensive and widely available. The one pictured in this book measures 60 × 40 × 30 centimeters.

- An electric heating mat: These are used to sprout greens in nurseries (look for a "seedling heat mat") and also to warm reptile terrariums ("reptile heating pad"). They consist of a resistive electric coil running through a thick plastic cover, and provide even heat over a large surface area. You can find them in many sizes, and they're usually waterproof and easy to clean.

- A temperature controller: As with the speed-rack setup, this will act as your thermostat, adjusting the internal temperature of the fermentation chamber. Many models are equipped with a little hole for a screw so you can conveniently attach it to the outside of your box.

- A small humidifier (when making koji): The smaller you can find, the better. Plus, a simple hygrometer, an instrument used to gauge humidity; it will look a bit like an oven thermometer. Alternatively, you could use a humidistat, which functions much like a thermostat. While slightly more expensive, it will simplify things by regulating the humidity in the chamber for you.

- A trivet, or a few screws: In most cases, you want to keep your ferments elevated off the bottom of the cooler. A trivet will do the trick, but for better airflow, procure four screws that are long and sturdy enough to make it through the walls of the cooler and support the weight of a tray laden with ingredients.

47

Building a Fermentation Chamber
Out of a Styrofoam Cooler

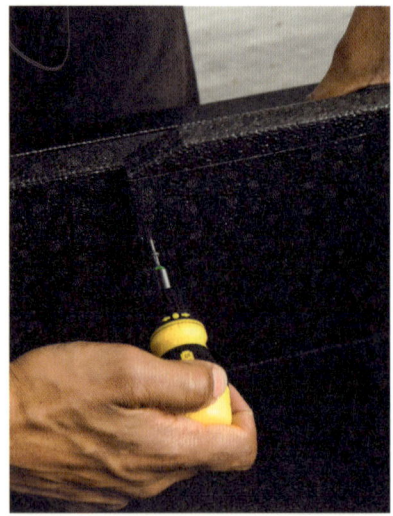

1. Ensure that your Styrofoam cooler is cleaned and sanitized. If you're making koji, procure four screws that are long enough and sturdy enough to bear the weight of a tray of koji and screw them into the sides of the container, about halfway up the walls.

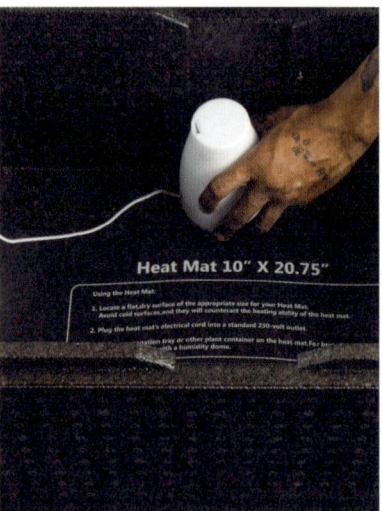

2. Place the heating mat and humidifier inside the container. Try to keep the humidifier off the heating mat, and snake the cords out of the box. Set the humidifier to medium and turn it on. Place your hygrometer (if you have one) next to the humidifier (out of the direct flow of steam) to keep track of the humidity.

3. Plug the heating mat into your temperature controller and, following the manufacturer's instructions, set it to your desired target temperature; for the ferments in this book, that will be either 30°C/86°F or 60°C/140°F. Run the temperature probe into the chamber.

4. Bring your chamber up to the desired temperature and humidity. If you don't have a humidistat, you'll adjust humidity by checking the level on your hygrometer and then dialing the humidifier setting up or down. The temperature controller will take care of temperature for you.

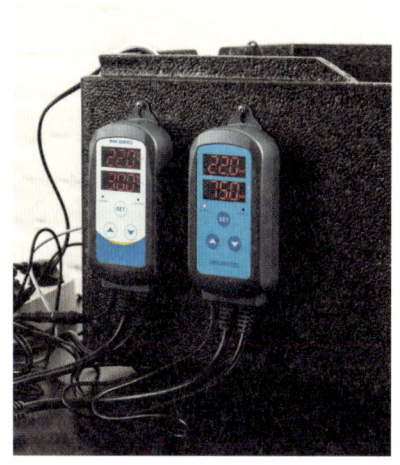

5. Add your ferment(s). Keep an eye on the temperature controller to make sure it's turning the heater on and off when the temperature dips or rises. You may see a drift of a degree or two above or below your desired temperature; that's normal.

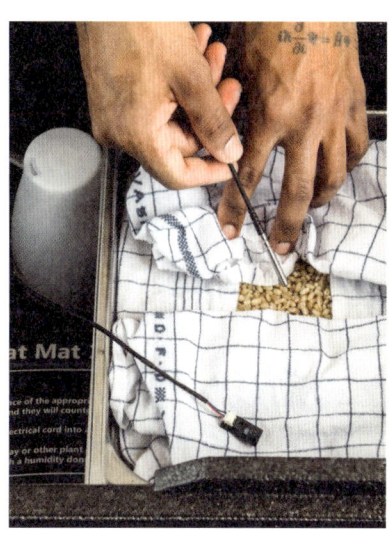

6. Cover the fermentation chamber with its lid. For ferments at 60°C/140°F, close it as tightly as you can to keep heat in. For koji, leave it cracked open on one side just a touch to allow fresh oxygen in. You can easily prop it open more with a screw placed into the lip if you're worried about it being closed shut.

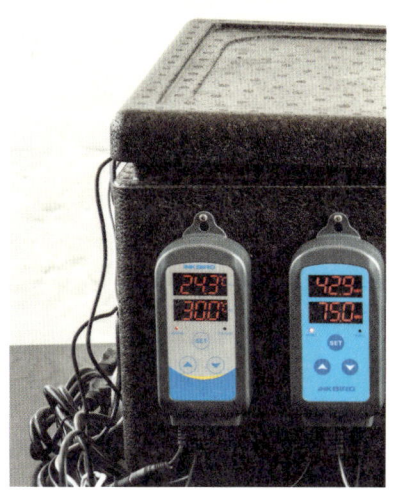

Thinking Outside
the Kraut

Our hope is that once you've read the text in each chapter
and made one or two of the corresponding ferment recipes,
you'll feel comfortable enough to begin steering the ship
yourself. We encourage you to take what you've learned and
apply it to other ingredients. One of the things we try to do
through our study of fermentation at Noma is to separate
techniques from their cultural framework to see what happens
when the biological processes are applied to different ingredients. It's not about disregarding the importance of cultural
history, but rather understanding how other culinary traditions
can improve the cuisine in our part of the world.

For instance, kimchi and sauerkraut are two of the world's
most well-known lacto-fermented products. That may be
obvious, but making the distinction between time-honored

foodstuffs and the techniques that produce them is an important step. Once you understand the role of a certain fermentation process—how it transforms ingredients, what it heightens and what it mutes—you can consider what else might benefit from the same treatment. What is it about cabbage that lends itself so well to being turned into sauerkraut? What other ingredients have similar qualities? What additional seasonings might complement the acidity brought by lacto-fermentation? That's how we direct a lot of the work in the fermentation lab at Noma, and it's led us to some of our most successful products.

Keep in mind that as you experiment, you'll inevitably fail. Don't get discouraged! Every recipe in this book began as an idea that edged its way to deliciousness through failure, education, and adjustment. Surprise and delight are only possible when things don't go according to plan.

Substituting Store-Bought Ferments

Our hope is that you'll come away from this book with a deeper understanding about the world of fermentation and cooking, even if you don't make a single ferment we've described. We want cooks and chefs everywhere to see the utility and value in fermented products, whether or not they make them from scratch. Shoyu isn't only for dipping, nor miso only for soup. If you come across a suggestion in this book that appeals to you, like, say, shoyu caramel, you shouldn't feel like you need to make your own shoyu to pull it off. Store-bought will do.

We also realize that some recipes in this book combine multiple ferments: sometimes out of necessity, other times to illustrate the powerful and flavorful interplay that can develop between different players. In such instances, you might have made one product but not its complement; a substitution will more than suffice to execute the recipe and get a good idea of the flavors we're chasing.

Unfortunately, we've never come across anything that bears a close enough resemblance to Maizo (page 312) or Grasshopper Garum (page 393) to recommend them, but here you'll find a chart of useful equivalencies for some of the products in this

51

Our ferment	Store-bought cousin
Elderberry Wine Balsamic (page 201)	Traditional balsamic vinegar
Pearl Barley Koji (page 231)	Dried rice koji
Yellow Peaso (page 289)	Okasan miso
Ryeso (page 307)	Hatcho miso
Yellow Pea Shoyu (page 338)	Raw shoyu
Beef Garum (page 373)	Worcestershire sauce
Rose and Shrimp Garum (page 381)	Fish sauce (Red Boat brand)

book—"cousins," so to speak. As always, quality counts. There will always be cheaper or more refined versions of products available on the market, and with fermented goods, the range can be quite drastic. Use your judgment and the advice of friends or grocery staff to determine which products are crafted with care and attention.

Weights and Measures

At Noma, and in this book, we use the metric system for all our measurements, because it allows for much greater precision and accuracy than imperial measurements. Whenever you're dealing with sensitive outcomes, accuracy is key. A shift in salt content of just 1 percent can be the difference between a ferment you'll want to show off to all your friends and something you'd rather no one ever knew about.

For any of our skeptical American readers, know that the metric system is supremely logical, and most kitchen measuring tools include metric markings and settings anyway. With the metric system, you can measure both weight (grams and kilograms) and volume (milliliters and liters). For many of our recipes, we use weight rather than volume for the sake of simplicity: Stick your empty bowl on a scale, tare it (meaning adjust the readout to zero, thereby discounting the weight of your bowl), and add the ingredient until you've reached the desired weight. No need to move ingredients between measuring cups and a work bowl.

A digital kitchen scale that measures to the single gram is essential to the execution of the recipes in this book. You can buy high-quality scales for not a lot of money; be sure to have an extra battery on hand so you aren't unexpectedly caught short in the middle of a recipe.

Finally, we've listed an approximate yield for each recipe so you'll know what you're getting into before you start, but it's easy to scale the recipes up or down. However, pay attention to the size of the container required. There are instances where a little extra headroom in a jar or crock might be desired, and if you scale up the recipe, you may also need to scale the container accordingly.

53

2.

Lacto-Fermented Fruits and Vegetables

—

Lacto Plums 69

Lacto Cep Mushrooms 83

Lacto Tomato Water 87

Lacto White Asparagus 93

Lacto Blueberries 97

Lacto Mango-Scented Honey 101

Lacto Green Gooseberries 105

Turning Sweet to Sour

There's not a dish on the menu at Noma, from the very first mouthful to the last, that doesn't involve some element of lactic acid fermentation (aka lacto-fermentation). Its usefulness is practically limitless.

Lacto-fermented products bring fruitiness, acidity, and umami to everything they touch. For instance, lacto-fermenting cep mushrooms (porcinis) yields an incredibly potent liquid that we use to season fresh sea urchins. Just a drop or two over each tongue of sea urchin will make your hair stand up—it invigorates and focuses the flavor of the urchin in an unbelievable way. It's like taking a picture of an urchin and cranking the saturation and contrast way up. As for the actual mushroom, we soak that in syrup, dry it, and dip it in chocolate to make a candy that accompanies coffee at the end of the meal.

Thankfully, lacto-fermentations are also incredibly straightforward to make. The process is simple: Weigh your ingredient, add 2% salt by weight, and wait. How many days depends on how sour you want the final product to be.

It's all made possible by the hard work of lactic acid bacteria, or *Lactobacillales* (we'll refer to them from here on out as LAB). LAB transform sugar into lactic acid, and they're the secret behind sour pickles and sauerkraut, rye breads and sourdoughs, yogurt, and sour beer. They're also involved (to a lesser extent) in making wine, cheese, and miso, contributing to the nuance and complexity of flavor that characterize these and many other iconic fermented foods.

Generally speaking, LAB are acid- and salt-tolerant, rod- and sphere-shaped bacteria. They're anaerobic, meaning they can flourish in the absence of oxygen. LAB consume carbohydrates, mostly in the form of sugars, and produce lactic acid as a metabolite (a by-product of their metabolism). Without getting neck-deep in the chemistry, the process involves the bacteria using enzymes to break down glucose ($C_6H_{12}O_6$) in order to harness its chemical potential energy, and thus converting each molecule of glucose into two molecules of lactic acid ($C_3H_6O_3$).

It's a microbe's world. We're just living in it.

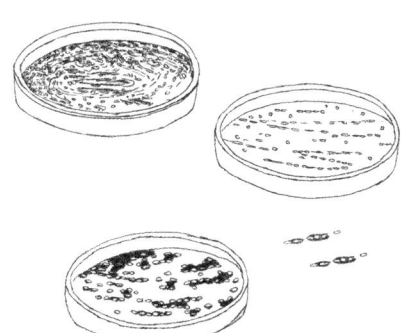

Different strains of lactic acid bacteria produce different flavors.

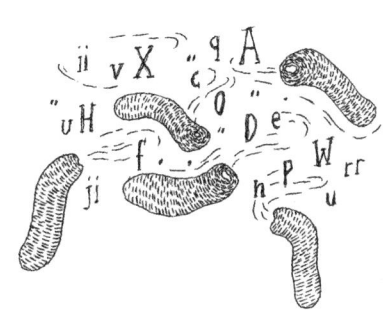

Bacteria can actually communicate with one another in a language of chemical gradients.

Species of LAB that specialize solely in converting sugar into lactic acid are classified as being homofermentative, while others are heterofermentative, meaning their metabolites can include not only lactic acid but also other molecules such as alcohol, carbon dioxide, or acetic acid. Some species of LAB are able to break down proteins into amino acids, giving cheeses like cheddar and Parmigiano their unspeakable deliciousness.

Like humans, LAB are industrious creatures that have managed to occupy environments the world over. They're present in the milk of mammals, meaning you've been involved in an intractable relationship with these bacteria since the first moments of your life. And fortunately for us, LAB are present on the skins and leaves of just about any vegetable or fruit you would ever want to ferment, patiently waiting for conditions to favor their needs.

At Noma, we practice "wild fermentation" for nearly all our lacto-fermented products, allowing the normal populations of bacteria already living on our food to kick-start the fermentation process. In any given wild ferment, there will be multiple strains of bacteria vying for position, blooming and fading at different times, each adding its unique voice to the chorus of flavors. It's the intricacy of this interaction between different LAB that makes wild ferments so delicious.

One of Noma's longtime friends, Patrick Johannson (aka the Butter Viking), once sent a sample of the wild cultured butter he'd made to a food lab for analysis, and found twelve distinct species of LAB cohabiting within it. Commercial operations often try to approximate the intricacy of wild ferments by manipulating factors like the temperature of a ferment over time, tweaking the conditions to suit different bacteria that produce specific flavors. LAB behave differently depending not only on temperature, but also the availability of nutrients, population density, and who their neighbors are. Chemical cues allow for communication between microbes that informs everything from their growth patterns to their rate of reproduction.

57

Beyond the Cucumber

The most common lacto-fermented vegetable in the Western world is the standard-issue sour cucumber pickle, which is lacto-fermented in brine. At Noma we look further afield for vegetables to lacto-ferment, but we always keep in mind the characteristics that make a basic dill pickle so enjoyable to eat. We look for things that are (1) tasty when raw and (2) juicy but not mushy. The latter trait is important because so much of a pickle's appeal is its crunch. (As any Scandinavian will tell you, slices of cured fish garnished with bits of pickled vegetables are one of life's great textural partnerships.) We've had amazing success making lacto-fermented pickles from white asparagus, small pumpkins, beets, and cabbage stems. Leafy greens like watercress and ramsons have been . . . less rewarding.

Of course, vegetable pickles are only one direction in which you can go. Once you understand that anything with sugar can be lacto-fermented, it opens up a world of potential. It's an absurdly basic realization, but once it occurs to you, you can't stop thinking, *What else can I give the lacto treatment?*

Every September at the restaurant, at the end of the berry season, we lacto-ferment blueberries, raspberries, mulberries, blackberries, white currants, and pretty much any other soft fruit we can get our hands on. Even though it lacks the crunch of a fermented root vegetable, the finished puree-like mash is a prize in itself—both sweet and savory, with multiple layers of sourness.

As LAB ferment sugar, the resultant lactic acid mingles with the acids already present in the fruit. Citric acid—most commonly associated with citrus fruits but also found in many other fruits and berries—can be quite tart and almost give off a burning sensation. Malic acid, found in grapes and apples (think of the tartness of a Granny Smith), is much rounder and mouthwatering. Ascorbic acid is sharp and direct, and can be found in all kinds of tropical fruits, from bananas to guava. The interplay of different acids is one of the most interesting and beautiful facets of fermented fruits.

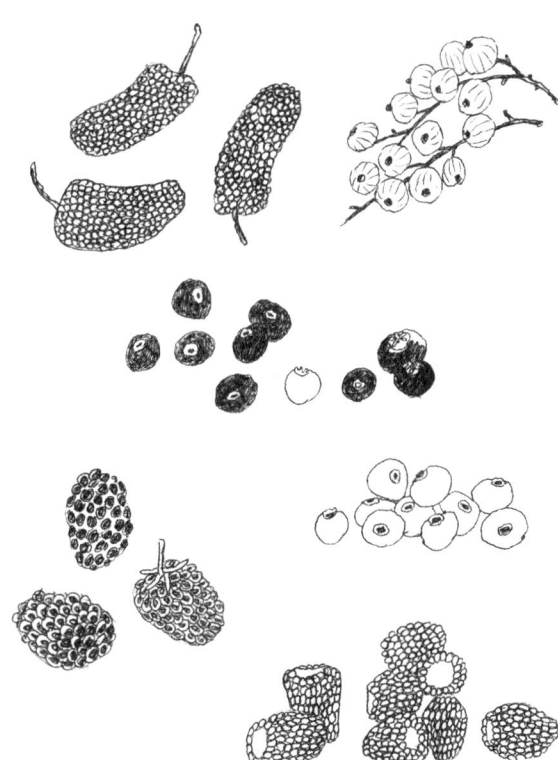

Lacto-fermented berries are powerhouses of flavor.

59

Because the berries usually lose their shape and texture when lacto-fermented, we often use a juice press to harvest the juices. Fermented berry juice is incredible—it has body and effervescence, saltiness, sweetness, and acidity. Mix fermented raspberry juice with a spicy olive oil, add a few grinds of floral spice—maybe long or pink peppercorns—and spoon the resulting vinaigrette over thick slices of ripe beefsteak tomatoes. Sprinkle it with sea salt, sugar, and a few torn leaves of marjoram, and it's the perfect distillation of late summer. And don't throw away the berry pulp. It will bring nuance and brightness to a bowl of fresh berries, topped with freshly whipped cream.

Putting
LAB to Work

As mentioned earlier, lacto-fermentation is gloriously uncomplicated, thanks in no small part to the fact that LAB can be found almost everywhere. That being said, there are a few basic conditions that must be met before LAB will perform at all (much like rock stars). Here are the various things you should do to ensure a successful lacto-fermentation.

Remove the Air

LAB function best in the absence of oxygen. With many traditional lacto-fermentations, liquid displacement is all it takes to keep LAB happily deprived of air. Take sauerkraut, for instance. Shredding the cabbage ruptures the plant's cells and releases moisture. Salt draws even more water out of the plant via osmosis, and a weight placed on top of the cabbage submerges it in its own juices, allowing the LAB to do their work.

At Noma, however, smushing vegetables under weights isn't always an option, so we'll go to great lengths to ensure that the finished fermented fruits or vegetables stay beautifully whole for their final presentation. We use plastic bags and a vacuum sealer to ensure that our LAB don't come into contact with oxygen.

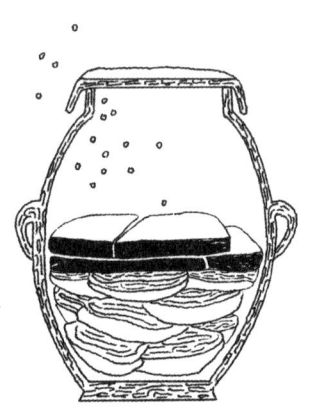

Packing jars tightly will help remove air and prevent spoilage.

However you choose to remove oxygen from LAB's environment, you're not only helping the bacteria to carry out their fermentation work, you're also excluding potential pathogens. By taking oxygen out of the equation, you also sabotage unwanted molds, which require air for their cellular respiration.

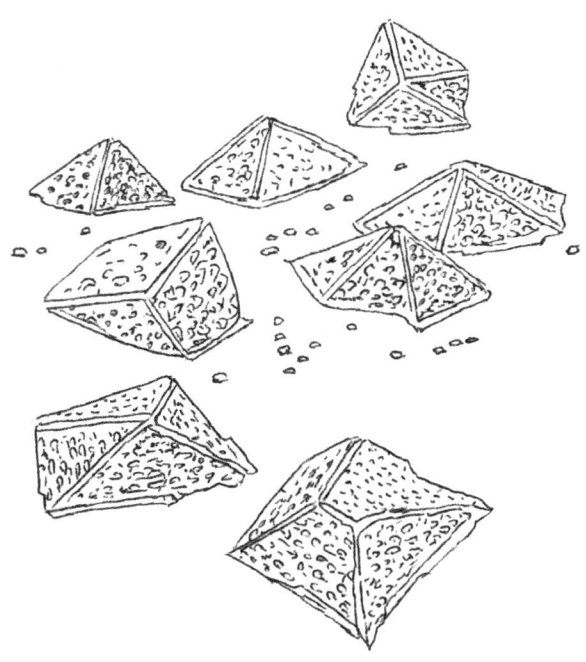

Salt Sufficiently

LAB don't require salt to flourish, but they do tolerate it, meaning we can use the salt content in a lacto-ferment as further insurance against unwanted outsiders. For instance, even though *Clostridium botulinum* is an anaerobe (a microbe that thrives in the absence of oxygen), it struggles in the presence of salt or acid, which is great news, because it's the bacteria responsible for botulism.

Different species of LAB exhibit different degrees of halotolerance (salt tolerance), with some species able to carry out their fermentation work in salt concentrations as high as 8% by weight. At Noma, we start our lacto-fermentations with 2% salt. It's enough to dissuade any malevolent bacteria from taking hold, but not so much that the product becomes unpalatably salty.

You can also create an environment that is both oxygen-free and salt-rich by fermenting in brine. Many traditional ferments, like the sour pickle, have been made this way for

61

centuries. Softer fruits will begin to dissolve in brine over the course of a few days, but crunchier vegetables of manageable size (beets, radishes, or young carrots) do extremely well submerged in salted water.

When lacto-fermenting in brine, start off by placing your empty crock or jar on a scale and taring the scale. Next, place your vegetables in the vessel, making sure they fit snugly without being squished. Cover the vegetables with enough water to fully submerge them, and note the total weight of the contents. Calculate 2% of that weight, and measure that much salt into a mixing bowl. Pour the water out of the vessel into the mixing bowl and blend until the salt is completely dissolved, then pour it back into the vessel. Note that the salt content in this method will always be higher than our standard of 2% salinity. For example: Assuming it takes about 1 kilogram of water to cover 1 kilogram of, say, cauliflower ribs, you'll add 40 grams of salt to the water to create a 4% salt brine. As time passes, salt will enter the fruit or vegetable and draw moisture out. With this ratio of brine to product-to-be-pickled, the 4% salt content will eventually even out, nearing 2% once the fermentation is complete, leaving you with perfect pickles.

If you have a pickle jar with a mildly tapered neck, it will help keep the vegetables from floating above the water line during the pickling process; otherwise, you can use some form of weight or barrier to keep the vegetables submerged. Leave a couple of centimeters of empty space at the top, and screw the cap on less than completely tight, preventing anything from entering but giving gas an easy escape route.

Choose Your Products Wisely (and Clean Them Lightly)

Avoid fruits or vegetables that have been coated in wax, treated with pesticides, or irradiated. Choosing organic is a good way to check off all those boxes. To ensure that you have a healthy population of wild LAB, avoid washing the products too thoroughly. If there's visible dirt, remove it by gently rinsing in cold water. No scrubbing, and no fruit or vegetable wash.

64

Chocolate from Native Jaguar Cocoa
and Mixe Chile, Noma Mexico, 2017

These pasilla mixe chiles are braised in
lacto mango-scented honey and stuffed
with chocolate sorbet.

Be sure not to ferment anything already sprouting mold or rotting. Fermentation is pretty magical, but it can't bring rotten apples back from the dead. Plus, you'll be starting out with unwanted microorganisms that can prevent LAB from thriving. That's not to say that you should be afraid of stretching the life of leftovers with lacto-fermentation. Chopping up a mixture of leftover strawberries and cherries and tossing them in a mason jar with some salt will yield a delicious topping for frozen yogurt a week hence.

Control the Climate

Most lacto-fermentations will function just fine at an ambient room temperature of about 21°C/70°F, but at Noma, we keep most of our lacto-fermentations in a room set to 28°C/82°F. We find this to be the ideal temperature for speedy fermentation, while avoiding *too* much bacterial activity, which can generate off-flavors. Lacto-fermentation will also take place in the refrigerator, albeit at a far slower pace.

One thing to note: If you want to prevent pickles from getting mushy, ferment them away from heat. There are natural enzymes in vegetables that will break them down more rapidly at higher temperatures. If you're especially concerned with maintaining pickle crunchiness, adding tannic plant leaves— like grape or horseradish leaves— to your brine or using mineral-rich unrefined sea salts or alum can reinforce the pectin in the plant walls and keep them snappy.

Consider Accessorizing

Due to the number of components in the recipes at Noma, we try to keep the flavors of our ferments relatively pure so they remain as versatile as possible. If we were to flavor our pickles with bay leaves, for instance, they could really only be used in places where bay makes sense. But that's not to say that you shouldn't season your concoctions during the fermentation process. Dried aromatics such as bay leaves and mustard seeds are obvious accompaniments for lots of sour ferments, but they aren't the only way. Try replacing 5 to 10 percent

65

of the water in a brine with fruit juice to inject brightness while supplying additional sugar for LAB to ferment. Vibrant fresh herbs like verbena or lemon balm can be infused into the brine beforehand or added as dried ingredients once the fermentation is complete. For spice, add a knob of horseradish or a halved chile. Even when fermenting sous vide, you can add complementary seasonings to the bag or jar, as long as you take their weight into account when adding the salt.

Different vegetables sharing the same brine can also exchange flavors. Cauliflower and salsify make great bedfellows. Onions and turnips lacto-fermented with a fistful of aromatic herbs like lemon thyme or orange blossom can elevate a dish of ceviche with floral notes and a crisp textural contrast. Use common sense when fermenting disparate ingredients together—don't pair blueberries with rutabaga and expect textural harmony— but one of the most wonderful and unpredictable aspects of fermentation is the way it draws new flavors from raw ingredients. What might be a pleasant pairing when eaten raw could end up being a mind-blowing combination once you've completed the complex dance with bacteria, salt, acid, and time.

Watch the Timing

It's important to catch a ferment at the right time. From the moment you place your fruit or vegetable into a salty environment, it begins moving in one general direction: from sweet to sour. While underfermented products can taste essentially raw, a ferment can easily be taken too far as well. Overfermented fruits or vegetables tend to have a sameness about them, where the original character and flavor of the product is washed away under a sea of sharp acidity.

Gauging when a fermentation is ready is no different from knowing when your pasta is perfectly al dente, or a floret of broccoli is properly blanched. As Thomas Keller once said, "Put it in your mouth and eat it." The only way to check on the progress of a lacto-fermentation is to taste it. An ideal lacto-fermentation should maintain the essence of the original raw product, but with added acidity, umami, and depth of flavor.

Fermentation is a practice in timing. It's up to you to decide when your ferment is "done."

Wait! Don't Throw That Away

This final point isn't crucial to a successful lacto-ferment, but it might help determine how successful you consider your project to be.

As we mention multiple times in this book, fermentation is a fantastic means of extending the life of food scraps that might otherwise go to waste. But the lacto-fermentation process itself can also create very useful by-products you might end up throwing away if you're not paying attention. Some of the world's most potent and delicious concoctions are leftovers from fermentation. Marmite and Vegemite are remnants of beer production. Sake lees, the residual rice pulp of sake production, are employed in numerous ways in Japanese cooking, most notably as a sweet-and-sour pickling agent for vegetables (*kasuzuke*).

Before you pour the pickle brine or the runoff from lacto-fermented plums down the drain, consider how it might taste in a soup or vinaigrette. Store it in sealed containers or repurposed condiment bottles in the fridge. If your lacto-fermented fruit or vegetable didn't turn out exactly as you had hoped, that salty-sour elixir can make for a fine consolation prize.

Halved plums, salted and ready to ferment.

Lacto Plums

**Makes 1 kilogram lacto plums
and juice**

1 kilogram firm-ripe plums
Non-iodized salt

Lacto-fermentation is a nice, gentle entrance into the world of fermentation—a place to get your foothold before you dive deeper into more involved projects. The process is straightforward and quick, and the rewards generally come in less than a week.

Lacto plums are an excellent place to start, as plums are readily available in many places and you can ferment them in a variety of ways, depending on what equipment you have and how you intend to use the final product—whether you're looking for chunks and pieces, whole plums, or a puree.

Equipment Notes

There are two ways to go about lacto-fermentation: You can ferment the raw product in either a vacuum-sealed plastic bag or in a vessel under weights. Vacuum sealers make lacto-fermentation supremely easy and consistent. They're a bit of an investment, but also incredibly useful for recipes throughout this book. On the other hand, you'll do fine with a tried-and-true glass jar or ceramic crock. You'll need weights of some sort in order to submerge the plums in the liquid that leaches out of them. Small ceramic or glass fermentation weights are great, but will be hard to find for smaller vessels. Zip-top bags filled with water will fit any container and work just as well.

It should also be mentioned that there are many points in this book where we advise you to wear gloves to prevent

Ripe but firm plums

Day 1

Day 2

Day 3

Day 4

Day 5

Day 6

Day 7

You'll probably notice that all the ferments in this book are depicted in glass jars and clear containers—we've done this so that you can see what's going on inside. However, you should note that prolonged exposure to UV rays can affect a ferment's health. Fermenting in a glass jar in direct sunlight may end up killing the beneficial microbes within. Exposure to indirect light, like, say, somewhere in your kitchen away from the window, is absolutely fine.

Time-Lapse Photos

Throughout this book, we illustrate the fermentation processes in a series of time-lapse photos to offer as much visual information as possible. In some instances, you may not notice huge differences from day to day or week to week, but we think you'll find it instructive to see the subtlety of the changes from photo to photo.

contaminating a ferment with competing microbes. When lacto-fermenting, it's important that your hands be washed and clean, but there's really no need for gloves. LAB are present everywhere, including your skin. By touching the food, you'll actually be adding a little bit of your own personal *terroir* into the mix.

In-Depth Instructions

Look for ripe but firm plums that are sweet and have a light crunch when you eat them raw. Underripe plums won't provide enough sugar for the LAB, leaving you with half-fermented fruit that's not sweet enough to balance the lactic acid. Overripe plums will disintegrate.

If the plums are visibly dirty, rinse them under cold water, but don't scrub them. The wild bacteria on the fruit's skin are the agents responsible for a successful ferment. Halve the plums lengthwise with a paring knife. Gently twist to pull the two halves apart, then wedge the blade of your knife under the tip of the pit and carefully lift it out; if the pit is stubborn, you may have to cut it out.

Weigh the pitted plums and calculate 2% of that weight—this will be the amount of salt you'll add later. For example, if the weight of your pitted plums is 950 grams, you'll need 19 grams of salt.

From here, depending on the equipment you have, you can proceed in one of two ways.

If fermenting in a vacuum bag: Place the plum halves in a vacuum bag large enough to accommodate all the fruit in a single layer. Add the salt to the bag, hold it shut by scrunching the top, and gently toss the contents around to distribute the salt evenly.

Lay the bag on a flat work surface. Reach in and use your fingers to arrange the plums in neat rows, cut-side down. With the plums still lying flat, turn on the vacuum sealer and seal the

Lacto Plums, day 1 (in a jar)

Day 4

Day 7

bag on maximum suction as close to the opening as possible. This will give you headroom, which will come in handy later, as you'll need to open and reseal the bag multiple times.

If fermenting in a jar or crock: Cut the plums in half two more times, into 8 pieces each. This will allow the plums to fit more snugly into the fermentation vessel and eliminate air gaps between pieces. Place the fruit in a bowl, add the salt, and mix well. Use a rubber spatula to scrape the plums and salt into your fermentation vessel of choice—make sure to get every bit of juice and salt.

Weight the plums down, so that as they release their juices, the liquid will cover the fruit in a salty brine. The easiest way to do this is to use plastic zip-top bags: Partially fill a bag with water, push out the air, and seal it. For extra security (in case you spring a leak), seal that bag inside another one. Place the water-filled bag into the crock or jar and wiggle it around so it covers the top of the plums completely. Cover the crock or jar with a lid, but be sure it's not so securely fitted that gas can't escape. If using a hermetic jar like the one pictured, you'll want to remove the rubber gasket before closing.

Whichever method you chose, you can now set the sealed plums aside to ferment. A lacto-ferment will function fine at a room temperature of 21°C/70°F, though at Noma, we lacto-ferment in a room that is a bit warmer, 28°C/82°F. This temperature is warm enough to accelerate the ferment's progress yet not so warm that the ferment will develop undesired flavors that overactive fermentation can sometimes produce. It is also possible to achieve successful lactic fermentation in a refrigerator—around 4°C/40°F—but the process will take much, much longer, with an increased risk of the fruit breaking down and browning before a sufficient amount of lactic acid has been produced. All things considered, we strongly recommend you ferment the plums at room temperature or warmer.

At 28°C/82°F, our plums usually take 5 days to ferment to the ideal flavor. At 21°C/70°F, they might take 6 to 7 days, but

A small zip-top bag full of water serves as a perfect fermentation weight, if you can't find small glass or ceramic weights.

ultimately you need to let taste be your guide. As the plums ferment, heterofermentative bacteria will produce carbon dioxide. If you vacuum-sealed your plums, this will cause the bag to inflate like a balloon. If it puffs up to the point where it looks like it might burst, you'll need to "burp" it: Snip a corner of the bag to vent the gas, then reseal with the vacuum sealer (be careful not to let the juices get sucked out of the bag). Resealing will also speed up fermentation by compressing the plums and forcing the bacteria-rich juices back into the flesh of the fruit.

While you're burping the plums, take the opportunity to taste them and check on their progress. In fact, you'd ideally taste the fruit every day. This is obviously easier with a crock or a jar than a vacuum-sealed bag, but if you left enough room at the top of the vacuum bag, you should have no problem opening and resealing it multiple times.

If fermenting in a crock or jar, look out for a wispy white substance that may form on the surface of the liquid and around the edges of your fruit. This is kahm yeast, a topical fungal bloom that can flourish before your fruit has fully fermented and acidified its juices. Kahm yeast is harmless, but it can add an off-flavor if it gets disturbed and mixed into the liquid. When you spot some kahm yeast, carefully spoon it off and discard.

As the fruit ferments, its flesh will soften and the sweetness of the plums will begin to turn into a pleasant acidity that hits you gently on the sides and back of your tongue, causing you to salivate slightly. The longer you ferment the plums, the stronger the acidic flavor will become. If you take it too far, eventually you'll lose the character of the fruit and all you'll taste is an overpowering acidity. Tasting the fruit every day will help ensure that you don't let the fruit overferment. Finally, note that lacto-ferments can have a slight fizzy quality due to the carbon dioxide produced by LAB being dissolved in the fruit's flesh—this is perfectly fine.

Once the plums have finished fermenting, remove them from the bag or fermentation vessel, and strain the juice through a sieve into a small container or plastic bag. Depending on the ripeness of your plums, you should have about 125 milliliters juice. The juice is an amazing product that's already halfway to being a fantastic vinaigrette. Store it in the refrigerator for up to a week or airtight in the freezer for long-term storage.

To store the lacto plums themselves, place them in a covered container or a resealable bag. They'll keep in the refrigerator for up to a week without changing character much, but if you're not using them immediately, freezing will prevent them from fermenting further. Fermented fruits keep much better in the freezer than fresh ones. If you kept your plums in halves, you can lay them cut-side down on a parchment-lined tray and freeze them solid, then place them in a vacuum bag, seal it, and return them to the freezer (a process known as the IQF, or individually quick frozen, method). Vacuum sealing is best for preventing freezer burn, but a regular freezer bag will work, too.

Finished lacto-fermented plums: sweet, sour, salty, and fruity.

1. Plums and salt.

2. Halve the plums with a paring knife.

3. Carefully remove the pits and discard.

76

4. Weigh the halved, pitted plums, then mix with 2% of their weight in salt.

5. Vacuum-seal the bag on maximum suction, leaving an ample amount of headroom.

6. Let the plums ferment for 5 to 7 days, or until the taste is to your liking.

7. "Burp" the bag by cutting a small hole in a corner and releasing the air. Taste the fruit to check your progress, then reseal the bag.

8. Somewhere between 5 and 7 days, the plums should be ready. Strain the liquid and reserve.

9. Transfer the plums to a sealed container and store in the refrigerator, or freeze them in a single layer to maintain their shape.

The skins of lacto-fermented plums can be dried, then ground into a tart and savory spice.

Suggested Uses

Chewy, Dried Lacto Plums

Drying the flesh of lacto-fermented plums gives them a welcome chewiness and enhanced savoriness that makes them even more versatile. Place peeled lacto plums—halves work best—on a parchment-lined baking sheet or the rack of a dehydrator and dry as close to 40°C/104°F as you can. You're looking to emulate the texture of dried apricots.

When it comes to putting the dried plums into action, think of them as less funky, fruitier dried preserved anchovies. Brown a stick of butter in a pan and toss in a small handful of torn sage leaves, along with a couple of spoonfuls of sliced dried plums. Add some crumbled, rendered fennel sausage, and fold in a batch of cooked pasta for a simple meal. Or use the same butter-plum-sage combination to baste pan-roasted cauliflower florets or spears of grilled white asparagus.

Plum Skin Chips

The skins of lacto-fermented plums can be slowly dried into a crisp chip in a dehydrator or a low oven. We dry our plum skins at about 40°C/104°F in a dehydrator. If you're using your oven, heat it to 60°C/140°F, if it will go that low. Dry the skins in a single layer on the dehydrator rack or on a parchment-lined baking sheet. The amount of time the skins take to dry will depend on your equipment, but the skins should have a nice snap that will improve as they cool. Dried plum skins can be sprinkled over anything from salad to brownies to ice cream for a punch of fruity acidity and a nice textural contrast.

79

Lacto Plum Powder

Use a spice grinder to pulverize dried lacto plum skins into a fine powder. The next time you grill a steak, rub it with a clove of garlic while it rests. Sprinkle a pinch of plum-skin powder and a few turns of freshly milled black pepper on top. The powder will melt into the crust and add a caper-like sharpness that cuts through the richness of good beef.

Making a fresh pea risotto for dinner? Instead of finishing it with a squeeze of lemon, just sift some plum-skin powder through a fine sieve over the top. This powder also cozies up nicely next to North African flavors—dust plum-skin powder over a platter of eggplant and chermoula just before serving to liven up an already dynamic dish.

Lacto Plum Juice Mignonette

The juice from lacto-fermented plums is amazing as a tart, salty wash for fresh seafood—it's outstanding as a direct replacement for mignonette, in fact. Next time you crack open some oysters, serve them with a ramekin of fermented plum juice rather than lemon wedges or Champagne-vinegar dressing. A half teaspoon over each oyster is like hitting "play" on the stereo.

Plum Custard

Lacto-fermented plum juice has intriguing sweet applications, too. For example, try making custard tarts flavored with the fermented juice. Bring 100 grams cream and 100 grams whole milk to a simmer in a saucepan. Meanwhile, beat 5 egg yolks with 50 grams sugar until pale, then add 75 grams fermented plum juice. Once the milk and cream have come to temperature, temper the egg mixture with a few spoonfuls of the hot liquid, then whisk in the rest until thoroughly combined. Strain through a sieve, then fill individual tart shells with the custard. Bake them at 170°C/340°F until set and then cool to room temperature for a slightly sour, salty-sweet take on a classic pastry. Double down on the plum flavor by dusting the finished tarts with plum-skin powder.

A dose of lacto-fermented plum juice
brings brightness and depth to a custard
of eggs, cream, and milk.

Above: Lacto-fermenting cep mushrooms
yields two products—the mushrooms themselves
and their incredible juice.
Opposite: Wild ceps are best foraged in the late
summer in the Northern Hemisphere.

Lacto Cep Mushrooms

Makes 1 kilogram lacto mushrooms and juice

1 kilogram cleaned cep (porcini) mushrooms, frozen for at least 24 hours
20 grams non-iodized salt

The true prize of this recipe is the fermented juice that leaches out of the ceps (aka porcinis). It's like a Swiss Army knife for us in the Noma kitchen—we use it to season everything from fennel tea to monkfish liver. It has a balance and funk that really electrify anything it touches.

In order to maximize the amount of juice we can harvest, we rupture the structure of the mushrooms' cells by freezing them before fermentation. That means prefrozen ceps are fair game for this recipe, as are fresh foraged ones. Oyster mushrooms, chanterelles, and bluefoots all ferment well and have their own distinct characteristics, if you can't find ceps. While less interesting, button mushrooms and creminis will work, too.

The in-depth instructions for Lacto Plums (page 69) serve as a template for all the lacto-fermentation recipes in this chapter. We recommend you read that recipe before starting in on this one.

If fermenting in a vacuum bag: Place the frozen mushrooms and salt in the vacuum bag and toss to mix the contents thoroughly. Arrange the mushrooms in a single layer, then seal the bag on maximum suction. Be sure to seal the bag as close to the opening as possible, leaving headroom that will allow you to cut open the bag to vent any gas that accumulates and then reseal it.

83

Lacto Cep Mushrooms, day 1
(vacuum-sealed)

Day 4

Day 7

If fermenting in a jar or crock: Mix the salt and mushrooms together in a bowl, then transfer them to the fermentation vessel, making sure to scrape all the salt from the bowl into the container, and press the mixture down with a weight. (A heavy-duty zip-top bag filled with water will do the trick.) Cover the jar or crock with a lid, but don't seal it so tightly that gas can't escape.

Ferment the mushrooms in a warm place until they have released much of their liquid, yellowed slightly, and soured nicely. This should take 5 to 6 days at 28°C/82°F, or a few days longer at room temperature, but you should start taste-testing after the first few days. If you're fermenting in a vacuum-sealed bag, you may also need to "burp" the bag if it balloons. (This should be less of a problem with mushrooms than other products.) Cut a corner open, release the gas, taste the mushrooms, and reseal the bag.

Once the mushrooms have reached your desired level of sourness and earthiness, carefully remove them from the bag or fermentation vessel. Strain the juice through a fine-mesh sieve. The mushrooms and their juice can be stored in separate containers in the refrigerator for a few days without a notice-able change in flavor. To prevent further fermentation, you can also freeze the mushrooms individually on a tray, transfer them to vacuum-sealed bags or zip-top freezer bags with the air removed, and store in the freezer.

The reserved juice can be clarified to produce a clear liquid that is potent with flavor. To clarify the juice, transfer it to a freezer-safe container with a lid and freeze. Once the juice is frozen solid, transfer the brick to a colander lined with cheesecloth and set it over a container to catch the liquid as it thaws. Cover with a lid or plastic wrap and place it in the fridge to thaw completely. Don't be tempted to wring out the cheesecloth once it's finished draining, as you'll end up forcing the mushroom particles through. Refreeze the clarified juice until needed.

Lacto Cep Mushrooms, day 1
(in a jar)

Day 4

Day 7

Suggested Uses

Candied Cep Mignardises

At Noma, we turn fermented mushrooms into dessert by soaking whole fermented ceps in their weight's worth of birch (or maple) syrup, then leaving them to infuse for 2 days in the fridge. Once they've become salty-sweet-sour, we dry them slowly in a dehydrator at 40°C/104°F until they have the chewy texture of toffee. Dip them in tempered chocolate, and they become sublime mignardises.

Cep-Bacon Vinaigrette

The juice from lacto-fermented ceps is a multipurpose seasoning tool we use often at Noma—there's a bright funkiness to it that electrifies certain ingredients. To get a sense for its powers, make this simple warm vinaigrette: Whisk together equal parts lacto-fermented cep juice and freshly rendered bacon fat. Spoon over grilled oyster mushrooms, slow-roasted cauliflower, or gooseneck barnacles.

Cep-Oil Companion

A perfect foil for lacto-fermented cep juice is cep oil. To make cep oil, heat 500 grams grapeseed oil and 250 grams fresh ceps in a saucepan over medium-low heat until the mushrooms begin to bubble. After about 10 minutes, cut the heat, cover, and allow the oil to cool to room temperature. Move the pot to the refrigerator and allow to infuse overnight. The following day, strain the oil and keep the confit mushrooms for another use. Whisk together equal parts cep oil and lacto-fermented cep juice, then stir in finely minced shallots or slivered garlic scapes, and you've got a sharp, savory dressing for raw scallops or lightly poached shrimp.

Above: Lacto-fermenting tomatoes doubles down on both their acidity and umami.
Opposite: Slowly straining lacto tomatoes separates the pulp from the tomato water.

Lacto Tomato Water

**Makes 1 kilogram lacto tomatoes
and tomato water**

1 kilogram ripe tomatoes
20 grams non-iodized salt

Tomatoes are already an acidic fruit with lots of umami, so the aim of lacto-fermenting them is not to sour them too much further, but rather to create a light balance of sour and sweet that almost fools you into thinking you're eating a cooked tomato sauce. As with so many of these lacto-ferments, the water from the lacto tomatoes is extremely useful in dressings and sauces. But that's not to say you should discard the pulp! Minced into a paste, it can be folded into lamb tartare, or spread on pieces of toast with fresh cheese, or blended with ricotta cheese for a lasagna filling.

The in-depth instructions for Lacto Plums (page 69) serve as a template for all the lactic fermentation recipes in this chapter. We recommend you read that recipe before starting in on this one.

Remove the stems from the tomatoes, and cut them into quarters if they're small or eighths if they're larger.

If fermenting in a vacuum bag: Place the tomatoes and salt in the vacuum bag and toss to mix the contents thoroughly. Arrange the tomato chunks in a single layer, then seal the bag on maximum suction. Be sure to seal the bag as close to the opening as possible, leaving headroom that will allow you to cut open the bag to vent any gas that accumulates and then reseal it.

87

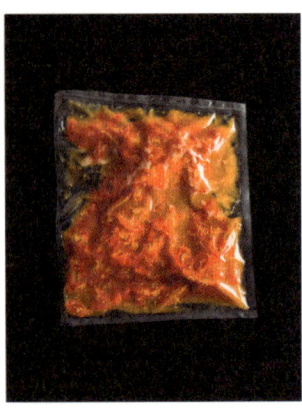

Lacto Tomato Water, day 1
(vacuum-sealed)

Day 4

Day 7

If fermenting in a jar or crock: Mix the salt and tomatoes together in a bowl, then transfer them to the fermentation vessel, making sure to scrape all the salt from the bowl into the container, and press the mixture down with a weight. (A heavy-duty zip-top bag filled with water will do the trick.) Cover the jar or crock with a lid, but don't seal it so tightly that gas can't escape.

Ferment the tomatoes in a warm place until they have released much of their liquid and softened considerably. This should take 4 to 5 days at 28°C/82°F, or a few days longer at room temperature, but you should start taste-testing after the first few days. If you're fermenting in a vacuum-sealed bag, you'll also need to "burp" the bag whenever it balloons up. Cut a corner open, release the gas, taste the tomatoes, and reseal the bag.

Once you're happy with the flavor of your tomatoes, line a fine-mesh sieve with cheesecloth and set it over a bowl. Pour the fermented tomatoes and their liquid into the sieve, wrap everything in plastic wrap, and leave it to drain overnight in the fridge. The next day, give the sieve a few sharp raps with your hand to coax out all the liquid, but don't press on the pulp.

The tomato water and the pulp can be stored in separate containers in the refrigerator for a few days without a noticeable change in flavor. To prevent further fermentation, you can also freeze them separately in vacuum-sealed bags or zip-top freezer bags with the air removed.

Lacto Tomato Water, day 1
(in a jar)

Day 4

Day 7

Suggested Uses

Lacto Tomatoes and Seafood

Almost any liquid harvested from a lactic ferment can be used to marinate or dress seafood—the water of lacto tomatoes being no exception. Season the lacto tomato water with your preferred chopped herb—try dill, chives, basil, or shiso—then finish by whisking in two spoonfuls of olive oil. Add a splash of shoyu for umami and potency, if you like, but the sauce will taste much fresher without it. Use the dressing to top raw oysters, clams, or slices of fresh bass or pike-perch.

Lacto-fermented liquids are not only for *finishing* seafood—they can be a potent cooking medium, too. Try steaming mussels in lacto tomato water, replacing the usual white wine directly with sour-salty lacto tomato water.

Tomato-Water Pickles

Lacto tomato water has all the acidity you need to cure a batch of quick, fresh pickles. Next time you're having friends over for a barbecue or garden party, try slicing up a few of your favorite crunchy vegetables—carrots, radishes, celery—and covering them with lacto tomato water. Add a sprinkle of salt and let marinate in the fridge for a minimum of 2 hours, or preferably overnight. Drain and set out the lightly pickled vegetables for your guests to munch on while you make dinner. If you can turn your kids on to fresh pickles like these, you'll have an easy snack on hand all the time.

Tomato Sauce

You can think of the pulp of lacto-fermented tomatoes as a replacement for store-bought tomato sauce. It's saltier and sourer, but it has richness as well. Next time you make a *ragù Bolognese*, replace one-quarter of the volume of your crushed tomatoes with lacto-fermented tomato pulp. If you find it too acidic, a spoonful of honey will help balance it out.

Or toast slices of country loaf on each side in a pan with good olive oil and season with salt. While the bread is still warm, spread a big spoonful of lacto tomato pulp over the bread. It's a flawless bruschetta as is, but if you like, you can garnish with torn leaves of basil or shavings of Parmigiano. If you really want to gild the lily, drape a thin slice of ham over the whole thing.

Lacto Tomato Leather

You can dry the pulp of lacto-fermented tomatoes into a fantastic chewy snack. Blend the pulp—seeds, too, as they add pectin, which provides body to the leather—on high speed until smooth. Pass the puree through a fine-mesh sieve and spread it in a thin layer on a sheet pan lined with a silicone mat. Dry in a low oven (around 50°C/122°F) until leathery and allow to cool before peeling it away in sheets; you can also use a dehydrator for this. It's a treat to eat as is or brushed with a bit of honey.

Lacto tomato water mixed with fresh dill makes
a savory, herbaceous dressing for seafood.

Lacto-fermented white asparagus has
an ideal crunch, balanced bitterness, acidity,
and umami.

92

Lacto White
Asparagus

Makes 500 grams

Water
Non-iodized salt
500 grams trimmed white asparagus
½ lemon, cut into 0.5-centimeter/
   ¼-inch slices

White asparagus is a treat we look forward to eating each spring, but its season is short. Almost as soon as it shows up in the kitchen, it's gone. Fermenting the asparagus gives it an afterlife that keeps us sustained through the colder months of the year.

This recipe was taught to us by our longtime friend and anarchist farmer Søren Wiuff. The mild bitterness of the asparagus interacts with the citric acid of the lemon and the lactic acid formed during fermentation to create a harmony not unlike that of a perfectly ripe grapefruit. Slice the lacto asparagus in half lengthwise and serve it as an accompaniment to charcuterie, or slice the stalks crosswise into thin coins for a bright and crunchy addition to just about any salad.

At Noma, we prefer white asparagus for its delicate flavor, but green asparagus ferments well, too.

The amount of salt and water needed in this recipe will depend on the size of the vessel you use. For 500 grams asparagus, a 2-liter mason jar should be a good size. In order to determine the right amount of salt and water, first place the jar on a scale and tare it (meaning adjust the readout to zero to discount the weight of the container). Stand the asparagus spears upright in the jar—they should be packed relatively tightly. Pour in enough water to cover the asparagus, and take note of the total weight of the water and the asparagus.

Lacto White Asparagus, day 1

Day 7

Day 14

Calculate 3% of that weight, and add that much salt to a mixing bowl. Pour the water out of the jar into the bowl. Whisk together the salt and water until the salt has dissolved. Pour the brine into the jar over the asparagus and distribute the lemon slices on top of that. Keep the asparagus spears submerged beneath the level of the brine by weighting them down with a water-filled zip-top plastic bag, a fermentation weight, or some other clean object wedged beneath the neck of the jar or crock. Cover the jar or crock with a lid, but don't seal it so tightly that gas can't escape.

Ferment the asparagus spears in a warm place (about 21°C/70°F) for 2 weeks. Begin checking them after a couple of days. If you taste lightly sour notes—beyond the lemon—you're on the right track. Once the asparagus has pickled to your liking, leave it in the brine, seal the container, and transfer it to the refrigerator. It will keep in the brine for several months.

Suggested Use

The New Gherkins

We like to deploy lacto-fermented asparagus spears the same way you would gherkins, as refreshing palate cleansers or tart garnishes. Serve them simply doused in a bit of olive oil during supper, whether you're having lasagna or grilled ribs. Or, the next time you're making burgers, thinly slice a spear of lacto-fermented asparagus and shingle the slices on one side of the cooked patty, then continue with your usual condiments. You'll be counting down the days until white asparagus comes into season again.

Pack the jar tightly but not so full that you bruise the asparagus.

Sprinkle salt evenly over the blueberries so
you don't damage the product with too much mixing.
If there are pockets of unsalted berries, they
won't ferment properly.

Lacto Blueberries

Makes 1 kilogram

1 kilogram blueberries
20 grams non-iodized salt

Lacto-fermented blueberries are one of the easiest and more versatile products in this chapter. They need no prep other than a quick rinse, and once they're done you'll find heaps of simple uses for them: Toss some onto your morning yogurt and granola, or add them to a smoothie, or puree the fruit and juices to make a salty-sweet coulis to be drizzled over ice cream or fresh cheese. Fermented blueberries freeze well and thaw quickly, making them easy to keep on hand at all times.

The in-depth instructions for Lacto Plums (page 69) serve as a template for all the lactic fermentation recipes in this chapter. We recommend you read that recipe before starting in on this one.

If fermenting in a vacuum bag: Place the blueberries and salt in the vacuum bag and toss to mix the contents thoroughly. Do your best to arrange the berries in a single layer, then seal the bag on maximum suction. If you're gentle with them, the blueberries will retain their shape through the fermentation. Be sure to seal the bag as close to the opening as possible, leaving headroom that will allow you to cut open the bag to vent any gas that accumulates and then reseal it.

If fermenting in a jar or crock: Mix the salt and blueberries together in a bowl, then transfer them to the fermentation vessel, making sure to scrape all the salt from the bowl into the container, and press the mixture down with a weight.

97

Lacto Blueberries, day 1
(vacuum-sealed)

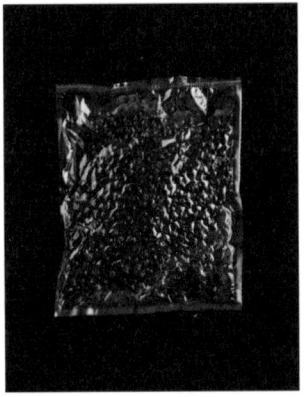

Day 4

Day 7

(A heavy-duty zip-top bag filled with water will do the trick.) Cover the jar or crock with a lid, but don't seal it so tightly that gas can't escape.

Ferment the blueberries in a warm place until they have soured slightly but still have their sweet, fruity perfume. This should take 4 to 5 days at 28°C/82°F, or a few days longer at room temperature, but you should start taste-testing after the first few days. If you're fermenting in a vacuum-sealed bag, you'll also need to "burp" the bag whenever it balloons up. Cut a corner open, release the gas, taste the blueberries, and reseal the bag.

Once the blueberries have reached your desired level of sourness, carefully remove them from the bag or fermentation vessel, and strain the juice through a fine-mesh sieve. The blueberries and their juice can be stored in separate containers in the refrigerator for a few days without a noticeable change in flavor. To prevent further fermentation, you can also freeze them separately in vacuum-sealed bags or zip-top freezer bags with the air removed.

Suggested Uses

Breakfast Topping

Fermented blueberries play a big part in our savory kitchen at Noma, but of course, most people think of blueberries as a sweet treat or a topping for yogurt in the morning. Fermented blueberries boost a simple breakfast into more sophisticated territory. A big scoop of plain yogurt, a spoonful of lacto blueberries, and a drizzle of honey will easily get you through until lunch.

Lacto Blueberries, day 1
(in a jar)

Day 4

Day 7

Lacto Blueberry Seasoning Paste

The pulp of lacto-fermented blueberries, blended smooth and passed through a sieve, makes for a tart, savory condiment for vegetables and meat alike. It's spectacular brushed on fresh corn on the cob with a bit of butter, or tossed with roasted beets. Paint barbecued ribs or pork chops with lacto blueberry paste before or after grilling, or make a barbecue sauce by substituting it for tomato paste or ketchup in your favorite recipe.

Puree lacto blueberries and spread them over corn on the cob.

Tropical fruits and chiles are hard to come
by in Denmark, but we enjoy working with them
whenever we can.

100

Lacto Mango-Scented Honey

Makes 700 grams

375 grams water
20 grams non-iodized salt
375 grams honey
5 grams fresh chiles, sliced
250 grams diced mango,
    with the skin on

Honey is more or less inert, meaning it will never go bad in your cupboard, but it will also never ferment in its natural state. While it contains a robust population of bacteria and yeasts in stasis, their microbial activity is halted because honey's sugar content is simply too high. We can work around this problem by diluting the honey to a sugar level that is low enough to support LAB. We've employed this recipe whenever we've ventured into warmer climes; it's found its way into desserts on our pop-up menus in both Australia and Mexico.

The in-depth instructions for Lacto Plums (page 69) serve as a template for all the lactic fermentation recipes in this chapter. We recommend you read that recipe before starting in on this one.

Whisk together the water and salt until the salt has dissolved. Add the honey and whisk again until fully incorporated.

If fermenting in a vacuum bag: Transfer the honey mixture to a vacuum bag, along with the chiles and mango, and seal the bag on maximum suction. Be sure to seal the bag as close to the opening as possible, leaving headroom that will allow you to cut open the bag to vent any gas that accumulates and then reseal it. Massage the contents gently to distribute them evenly.

101

Lacto Mango-Scented Honey,
day 1 (vacuum-sealed)

Day 4

Day 7

If fermenting in a jar or crock: Transfer the honey mixture to the fermentation vessel, add the chiles and mango, and muddle the fruit by pressing on it lightly with a spoon or spatula. Lay a sheet of plastic wrap directly on top of the liquid, making sure it reaches all the way to the edges, and cover the jar or crock with a lid, but don't seal it so tightly that gas can't escape.

Ferment the honey in a warm place until it has soured slightly and taken on the heat and perfume of the chiles and mango. This should take 4 to 5 days at 28°C/82°F, or a few days longer at room temperature, but you should start taste-testing after the first few days. If you're fermenting in a vacuum-sealed bag, you'll also need to "burp" the bag whenever it balloons up. Cut a corner open, release the gas, taste the honey, and reseal the bag.

Once fermented to your taste, strain the honey through a fine-mesh sieve and discard the mango and chiles. Feel free to save the strained fruits for another application (like a spicy chutney). You can store the honey in the refrigerator for a few weeks, or freeze it in a vacuum-sealed bag or zip-top freezer bag with the air removed for longer storage.

Lacto Mango-Scented Honey,
day 1 (in a jar)

Day 4

Day 7

Suggested Uses

Sugar Replacement

The most obvious use for lacto-fermented honey is as a (more delicious, more interesting) sugar substitute. Lacto-fermented honey steps in more than adequately for the sugar in compotes and jams, helping to keep the original character of the fruit more intact. Lacto honey is also fantastic in tea or coffee, especially if you're making chilled drinks, where its flavor can really shine through.

Honey-Poached Pears

Lacto-fermented honey is a perfect pairing to almost any fruit, but especially pears. In a wide-bottomed pot, combine 500 milliliters lacto honey, 500 milliliters white wine, and a spoonful each of chopped fresh rosemary and thyme. Lower 6 ripe but firm peeled pears into the liquid, cover, and bring to a simmer. Cook the fruit until softened slightly, 3 to 5 minutes, then remove the pot from the heat and allow to cool to room temperature. Not many people can resist these gems, sliced and served with a healthy scoop of ice cream, or dressed with a drop of vinegar and paired with hard cheese.

Lacto gooseberries were the sour spark that began
the pursuit of fermentation at Noma.

104

Lacto Green Gooseberries

Makes 1 kilogram lacto gooseberries and juice

1 kilogram firm-ripe green gooseberries
20 grams non-iodized salt

Gooseberries are beloved in northern Europe, but they grow in temperate climates around the world. Cultivars range in color from a jewel-like pale green to crimson red, with a striped pattern that runs longitudinally. For this ferment, we use green gooseberries that are just less than ripe and still firm if pinched. This recipe also works brilliantly with red gooseberries, which are usually juicier and sweeter than their green counterparts, and are quicker to ferment.

If fermenting in a vacuum bag: Place the gooseberries and salt in the vacuum bag and toss to mix the contents thoroughly. Do your best to arrange the berries in a single layer, then seal the bag on maximum suction. If you're gentle with them, the gooseberries will retain their shape through the fermentation. Be sure to seal the bag as close to the opening as possible, leaving headroom that will allow you to cut open the bag to vent any gas that accumulates and then reseal it.

If fermenting in a jar or crock: Mix the salt and gooseberries together in a bowl, then transfer them to the fermentation vessel, making sure to scrape all the salt from the bowl into the container, and press the mixture down with a weight. (A heavy-duty zip-top bag filled with water will do the trick.) Cover the jar or crock with a lid, but don't seal it so tightly that gas can't escape.

Ferment the gooseberries in a warm place until they have soured to your liking. This should take 5 to 6 days at 28°C/82°F, or a

Lacto Green Gooseberries, day 1
(vacuum-sealed)

Day 4

Day 7

couple of days longer at room temperature. The berries should be tart and mildly briny by the end, but you should start taste-testing after the first few days to make sure they're on the right track. If you're fermenting in a vacuum-sealed bag, you'll also need to "burp" the bag whenever it balloons up. Cut a corner open, release the gas, taste the gooseberries, and reseal the bag.

Once the berries have reached your desired level of sourness, carefully remove them from the bag or fermentation vessel, and strain the liquid through a fine-mesh sieve. The gooseberries and their liquid can be stored in separate containers in the refrigerator for a few days without a noticeable change in flavor. To prevent further fermentation, you can also freeze them separately in vacuum-sealed bags or zip-top freezer bags with the air removed.

Suggested Uses

Gooseberry Relish

Simply sliced in half, these soured berries make for a refreshingly bright palate cleanser in between mouthfuls of richer dishes, like seafood chowder. But lacto-fermented gooseberries can also serve as the backbone of a spectacular relish. Mince 100 grams fermented gooseberry flesh and combine with 15 grams each chopped parsley and chopped tarragon, 1 finely minced clove garlic, and a liberal drizzle of olive oil. Season with salt, if needed. This topping begs to be smeared over braised short ribs, barbecued quail, or grilled vegetables like asparagus or young leeks.

Leche de Tigre

The juice from a bag or crock of fermented gooseberries is one of our all-time favorite seasonings at the restaurant. All on its own, the salty, sour, fruity liquid can turn slices of firm raw fish like snapper or sea bream into delectable ceviche. To take that idea to another level, you can make a proper *leche de tigre* (the Peruvian term for ceviche marinade). With a handheld

Lacto Green Gooseberries, day 1
(in a jar)

Day 4

Day 7

blender, blend 1 part by weight peeled raw rock shrimp with 3 parts lacto gooseberry juice, then pass the mixture through a sieve. Add as much finely minced shallot and minced habanero to the liquid as you like, and pour the mixture over sliced raw fish, allowing it to sit for about 5 minutes before enjoying it with some chopped fresh cilantro.

Buttermilk-Gooseberry Dressing

Fermented gooseberry seeds are tiny but delicious—they retain a pleasant bite even through the fermentation process and benefit from the flavor of lactic acid, too. To harvest them from the fruit, cut the gooseberry laterally and press the flesh gently against a cutting board until the seeds pop out. There aren't many seeds per gooseberry, but their spectacular combination of texture and tartness will make the work worthwhile. Mix the seeds with a spoonful of buttermilk and a couple of cracks of black pepper to make a topping that cuts through fat and electrifies whatever it touches. It can be used to garnish freshly shucked steamed clams, slices of raw amberjack, or blinis topped with fish roe and elderflower crème fraîche (see the recipe on page 142).

The seeds of lacto gooseberries are as good as the flesh and well worth the effort to harvest.

107

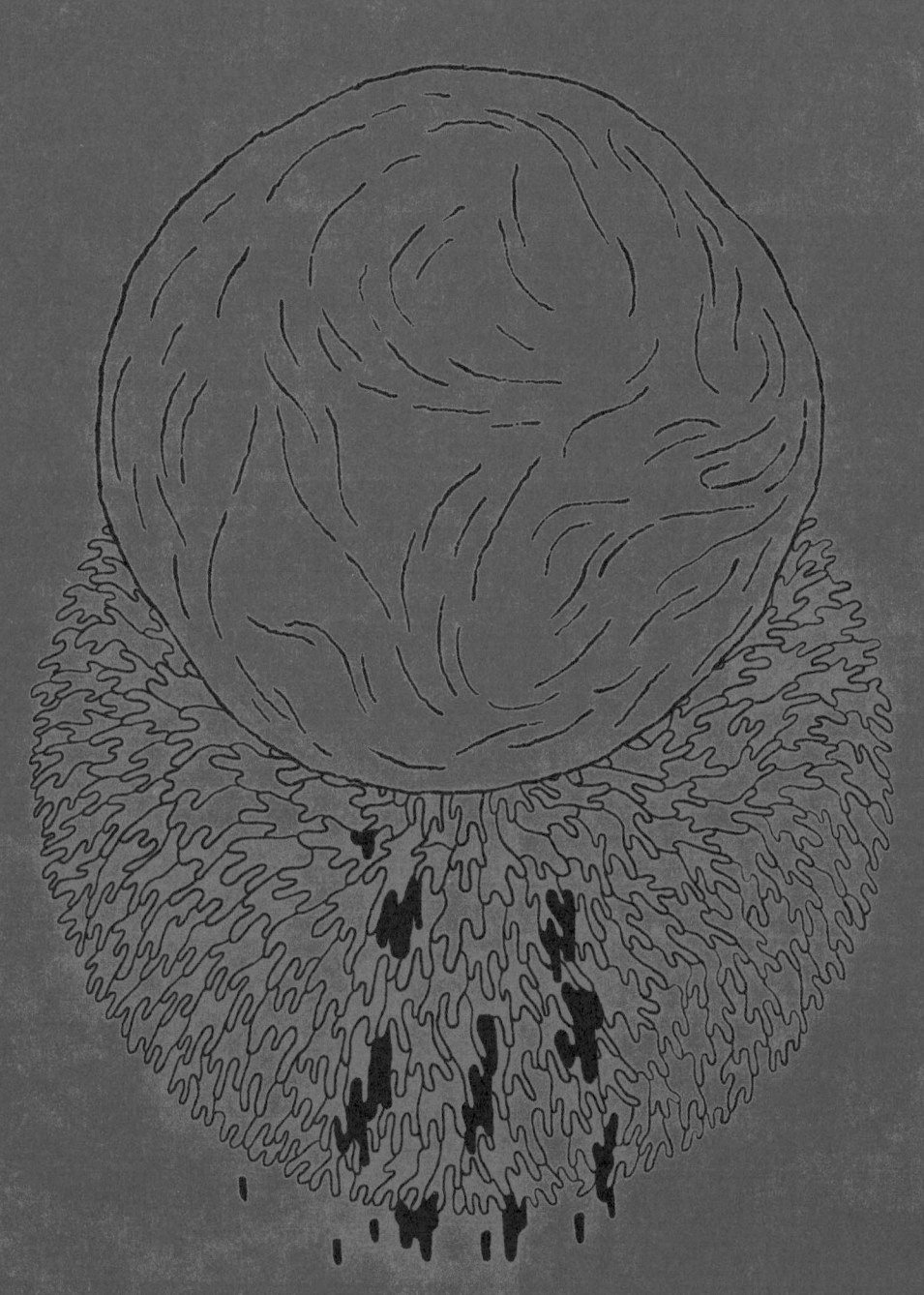

3.

Kombucha

—

Lemon Verbena Kombucha 123

Rose Kombucha 133

Apple Kombucha 137

Elderflower Kombucha 141

Coffee Kombucha 145

Maple Kombucha 149

Mango Kombucha 153

A Historic Brew
Reimagined

When we first began dedicating time and attention to learning about fermentation at Noma, we consumed every piece of relevant literature we could get our hands on. Each time we came across an unfamiliar term in a book, it gave us a thrill. The things we were reading about may have been centuries old, but they were essentially new to our little corner of the world. Ten years ago, for instance, hardly anyone in Denmark was drinking kombucha. When we first tried making our own, we had to go to Christiania—Copenhagen's self-declared autonomous hippie neighborhood where tourists go to buy hash—to find supplies.

Kombucha is a soured and lightly carbonated fermented beverage, traditionally made from sweetened tea. Its murky origins are believed to lie in Manchuria (what is now northeastern China) around 200 BCE. From there it spread east to Japan, largely through the efforts of a fabled Korean physician named Kombu. Hence, *kombucha* (*cha* being Chinese for "tea").

Historically, kombucha has been consumed in Japan, Korea, Vietnam, China, and parts of eastern Russia. But in recent years, kombucha has exploded in popularity across North America and Western Europe, thanks to clever marketing and the public's growing fascination with all things probiotic.

At Noma, we brew kombuchas as effervescent and vibrant bases for some of the juice pairings that many of our diners choose in place of—or sometimes in addition to—wine pairings. Kombucha has been incredible for opening up the range of our drink service at Noma. In the same way that we might develop a dish for the menu, we mix kombuchas with fresh juices, spices, oils, and even insects to create elegantly balanced beverages.

Almost any liquid with enough sugar can be fermented into kombucha, so while it's customary to make kombucha from sweetened tea, some of the varieties that we enjoy most are made from tisanes (herbal infusions) or fruit juices, which yield a roundness and depth of flavor not found in tea. We've made great kombucha from infusions of chamomile, lemon

Kombucha was born in ancient China.

verbena, elderflower, saffron, and rose, as well as from the juice of apples, cherries, carrots, and asparagus.

The kind of kombucha we're chasing bears very little resemblance to the sour liquid people force themselves to drink because it's supposedly good for them. To be perfectly honest, we find the average store-bought tea-based kombucha a bit boring. The flavor of the tea usually fades into the background, and you're left with a one-note beverage that doesn't really take you on any sort of a journey.

One of our earliest forays was a carrot kombucha that opened our eyes to what kombucha can be. It was like a perfect soup all on its own, a cold broth that still had some carrot-y sweetness, but with a touch of acidity. It had transformed into something with new dimensions that complemented the original flavor without obscuring it. Since then, we've been on an ongoing mission to brew kombucha from as many different bases as we can. Some of our explorations have taken us off the beaten path into experiments with dairy, tree saps, and stocks made with spicy chiles.

We cook with kombucha, too. Once you stop viewing kombucha as simply a New Age health drink, a number of culinary possibilities open up. The longer you ferment kombucha, the more acidic it grows. After a while, it becomes a lively ingredient in marinades or vinaigrettes, or an intriguing substitute for white wine or Champagne in sauces. Or you can reduce kombucha in a pan into a magical syrup—the kind of sweet, tart topping you really want on your pancakes.

Kombucha is produced by a collective of microbes that works in sync to first turn sugar into alcohol, then alcohol into acetic acid (the same acid in vinegar). As the microbes perform their work, they form a visible raft that's commonly known as the kombucha "mother," but also sometimes confusingly referred to as the "kombucha." Technically speaking, it's called a SCOBY (Symbiotic Culture of Bacteria and Yeast), so to keep things clear, we'll use the terms *kombucha* for the finished product and *SCOBY* for the organisms/mother.

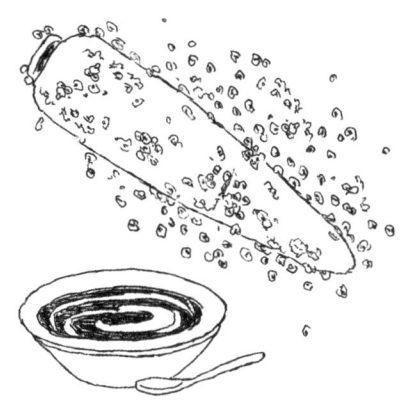

Carrot kombucha was one of Noma's first forays into kombucha brewing.

Cooperative Fermentation

A young SCOBY acquired from a fermentation shop.

And the same SCOBY after 7 days spent brewing a batch of kombucha.

The specific species of microbes that produce kombucha will vary from one place to another and from one batch to the next, but the main players are yeast (a unicellular fungus) and acetic acid bacteria (AAB, for short). The yeast is often *Saccharomyces cerevisiae*, but can also include many of its relatives. The AAB might also be a combination of several species, but some representative from the genus *Gluconacetobacter* or *Acetobacter* will always be present.

Once introduced to a sugary liquid, the yeast in a SCOBY kicks off a cascade of fermentation by consuming simple sugars and producing ethanol—the primary type of alcohol in wine, beer, and spirits—and a bit of carbon dioxide. The cohabiting AAB then ferment the ethanol by oxidizing it into acetic acid, utilizing the oxygen from its surroundings. The bacteria's quick work transforming alcohol into acid means kombucha doesn't have as much alcohol as wine or beer, but it isn't completely alcohol-free. The ABV (alcohol by volume) of kombucha hovers in the neighborhood of 0.5% to 1%. For context, a typical beer is around 5% ABV, and wines are in the low teens.

As a rough guide:

- Under ideal conditions, yeast typically ferments 2 units of sugar into 1 unit of alcohol.
- AAB convert 1 unit of alcohol into just under 1 unit of acetic acid.

The operative word in SCOBY is *symbiotic*, a term that usually implies harmonious cooperation but actually covers a multitude of different relationships. *Symbiosis* translates from Greek as "living together," and parasites, pathogens, and commensal organisms (those that benefit from another without harming it) all qualify. At one extreme, parasites weaken or even kill their hosts. On the other end, mutualist relationships provide benefits to both parties. AAB are commensal to yeast: The bacteria get more out of the relationship, but they don't harm the yeast. Yeast that cozies up to AAB is fairly tolerant of acidic environments and isn't bothered much by its acidic partners.

Bergamot Kombucha with Native Mint,
Noma Australia, 2016

A kombucha made from bergamot tea muddled
with native Australian mint and fresh citrus
for our juice pairing.

The bacteria and yeast in a SCOBY live together on a structure called a zoogleal mat (the aforementioned raft). As the bacteria in the SCOBY multiply and propagate, they excrete cellulose, forming a buoyant sheet that floats atop the liquid like a gloopy jellyfish. As the mixture ferments, the mat grows, spreading out on the surface of the liquid to the boundaries of the container, then increasing in thickness. Living on the mat allows the AAB to come into direct contact with the air above the liquid, which they require to convert alcohol to acid.

Vinegars are made sour in a very similar process, but with one important distinction: Vinegar is a two-stage fermentation. In the first stage, yeast ferments the sugar to alcohol. Different yeasts have different tolerances for the alcohol they produce, and will die off once that level is met (or they can be killed off by pasteurization whenever the vinegar maker chooses).

In the second stage, AAB ferment the alcohol into acid, but without the yeast, the bacteria will eventually run out of fuel and fermentation will stop. Choosing a strain of yeast that will die off on its own—or killing the yeast after it has produced just the right amount of alcohol—is how brewers control the process, and thus the sourness of their vinegar.

Kombucha, on the other hand, is a sustained fermentation. The yeast continually ferments sugar into ethanol for the bacteria to convert into acetic acid. That means that unlike vinegar, which can be aged for years or even decades while still remaining slightly sweet, a kombucha will become more and more sour until all the available sugar is used up. Even if you harvest a kombucha at the right moment and transfer the liquid to the fridge, it will continue to acidify.

That's why some commercial varieties of kombucha can taste overfermented and sour rather than refreshing. A well-made kombucha should have enough residual sweetness to be enticing and enough acidity to be lively. Striking the ideal balance between sugar and acid comes down to the third member of the kombucha symbiosis: humans. It's up to us to determine when a kombucha is ready to harvest.

Interestingly, the role that humans play also extends to the evolutionary history of the bacteria and yeast responsible for kombucha. Before the advent of microbiology, the best indicators of a SCOBY's vigor were visual cues. A robust SCOBY was likely taken as a good sign and prized by kombucha makers, who would save and propagate such specimens, thus giving precedence to the bacteria that were able to produce them. While the microbes in a SCOBY don't need a zoogleal mat to function, human intervention has ensured the survival of SCOBYs that produce thick rafts.

The Sweet Spot

Let's have a look at the basic process for making a kombucha:

1. First make a juice, tea, or infusion. Sweeten it and cool it.

2. Backslop with a previous kombucha to lower the pH. (See page 33.)

3. Add a SCOBY, or a cutting from one. A piece that covers at least 25 percent of the surface of the liquid should suffice.

4. Cover the container and leave the mixture to ferment, ideally at slightly warmer than room temperature.

5. Taste the kombucha often. Once it's reached your ideal balance of sweetness and acidity, remove the SCOBY and reserve it, then strain and chill the kombucha.

So, with all the talk of balance and timing in this chapter, a glaring question remains: How much sugar should you add when making kombucha?

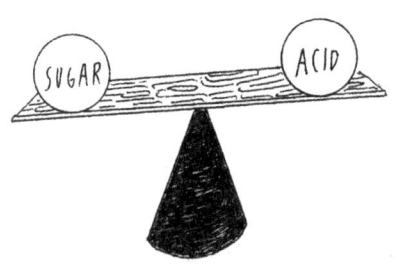

As a kombucha's sweetness diminishes, its acidity rises.

As a frame of reference, if you were to place a SCOBY in a jar of plain water, it would die, starved of the sugars needed to fuel its metabolism. Conversely, if you were to place the SCOBY into a tub of saturated sugar solution, it would also die, shocked by the high concentration of sucrose; the microbes simply wouldn't be able to function. (This is the same reason that honey, which is mostly made of sugar, never goes bad.)

In short, there's no perfect answer to how much sugar you need in your kombucha, but a good bit of trial and error have led us to a number we're happy with.

Sweetness is expressed in degrees Brix (°Bx)—a measure of the amount of dissolved sucrose (table sugar) as a percentage of the total solution (grams of sugar divided by the total grams of sugar and water combined). We find that a kombucha made from a high starting Brix of 35°Bx (very sweet) will not be anywhere near as delicious as one made from a more moderate starting Brix. A kombucha started at 35°Bx will contain too much sugar *and* too much acid. We've settled at 12°Bx for most of our kombucha recipes.

As for how long to let kombucha ferment, it's helpful to visualize the fermentation process as a curve. In the beginning, the liquid tastes familiar and somewhat lifeless. For example, a sweetened infusion of elderflower begins life tasting like flat soda pop, but by day seven, at the peak of the curve, it can resemble an amazing sparkling wine (sans alcohol). The original floral flavor will be clear, with added effervescence and brightness. But after that high point, the kombucha will steadily move down the other side of the curve toward an unpalatably acidic and shocking endpoint.

More so than with other ferments, pulling kombucha and consuming or cooking with it at the right time is critical. At Noma, we often freeze our kombuchas to halt fermentation and ensure that they're held at the top of their curve. We could also achieve the effect via pasteurization, but by applying heat you invariably muddle and mute the flavors of the ferment.

115

Caring for Your SCOBY

As delicious as we find kombuchas, we also think they're fascinating to work with and watch grow. As you're tending your own kombucha, you'll grow attached to it, in the way that people with sourdough starters come to delight in nurturing them.

While there's no kombucha pet store, there are plenty of websites where you can purchase live SCOBYs (see Sources, page 448), not to mention natural foods and home-brew shops. SCOBYs are normally sold vacuum-sealed or jarred in a small quantity of soured kombucha base. A healthy specimen should resemble a semi-opaque disk of firm gelatin. As a SCOBY requires air to thrive, it's important to transfer it to an open-air container as soon as you receive it.

If you aren't planning to brew kombucha immediately, you'll need to hold your SCOBY in stasis until you are. Whip up a batch of 20% sugar syrup (800 grams water and 200 grams sugar, brought to a boil, then cooled) and transfer it, along with your new SCOBY, to an open-top jar. Cover with a breathable towel or cheesecloth and secure it with a rubber band. Be sure to add any liquid that came with your SCOBY, as it's full of the same bacteria, yeast, and acid that make for a happy colony.

Now, as with a sourdough starter, a SCOBY needs a little upkeep between kombucha brewing sessions. If you make kombucha regularly, you can get into a steady rhythm of transferring a SCOBY from one batch to the next, always keeping it busy. But if you harvest your kombucha and aren't ready to start a new batch, you'll need to keep the SCOBY floating in roughly twice its weight of kombucha or sugar syrup. The liquid will sustain it for a while, but eventually the SCOBY will convert all the sugar into acid and you'll have to transfer it to a fresh home. Every 2 or 3 weeks, you'll need to repeat the process detailed above, brewing fresh syrup and transferring the SCOBY. (SCOBYs can be stored in the fridge to slow their metabolisms, but we prefer to keep them near room temperature so they're always ready for action.) If you notice the top of the SCOBY drying out, be sure to baste it with liquid from underneath, in order to keep its surface acidified.

Caring for your SCOBY means preparing a nice home for it in between batches of kombucha.

Another important factor in the longevity of your SCOBY is priming its environment (what we call backslopping; see page 33). If you place a SCOBY straight into sweetened liquid (tea, fruit juice, milk, etc.), there's likely to be wild yeasts and bacteria that can compete for the sugars within the liquid and impart musty, unpleasant flavors. Worse still are wild fungal molds, such as malevolent strains of *Aspergillus*, which can produce water-soluble toxins. To help prevent unwanted microbes from flourishing, you'll need to add some kombucha from a previous batch (or from a store-bought kombucha, if this is your first batch).

By adding some kombucha to the mix (10 percent of the total weight), you lower the pH of the solution, usually to below 5, which is enough to prevent invaders from blooming. The SCOBY, on the other hand, not only tolerates a low pH level but thrives in it. Priming also adds a booster shot of the bacteria and yeasts you're looking to propagate within your solution.

One thing to note is that SCOBYs can acquire the flavor of the base liquid that they're fermenting, and pungent flavors can carry over to the next batch. To avoid this, use SCOBYs to ferment the same or similar-flavored liquids each time. At Noma, we have a "living library" of SCOBYs for each type of liquid base.

SCOBYs are communities of organisms in which humans play a vital role.

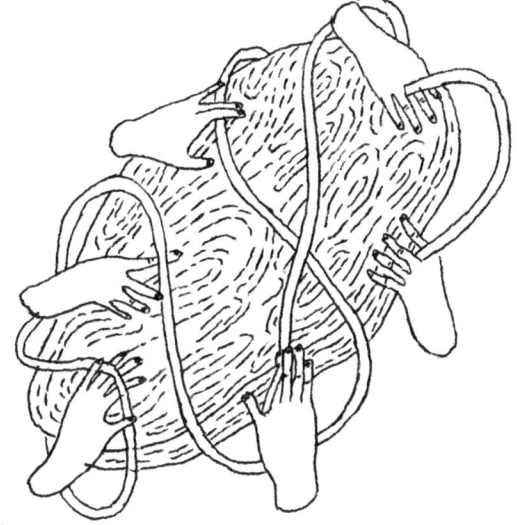

The Brix scale is named for Adolf Brix, a German engineer who devised the system in the early 1800s for use in the beer and wine industry. Degrees Brix (°Bx) aren't a measurement in themselves, but rather a scale that corresponds to the specific gravity of a solution. Specific gravity is the ratio of the density of a solution (such as a sugar syrup) to the density of normal water. The more sugar in the solution, the higher the specific gravity. The Brix scale translates specific gravity into degree increments, giving you a numerical evaluation of a liquid's sweetness.

For a sense of the scale: Simple syrup (1 part sugar and 1 part water by weight) has a measure of 50°Bx, while a double syrup (2 parts sugar and 1 part water by weight) has a measure of 66.7°Bx.

You can measure degrees Brix with a refractometer. Sucrose, when dissolved in a solution, changes the way water refracts light. A refractometer measures this change in angle and correlates it to degrees Brix. We use a refractometer at Noma to keep close tabs on our fermentation projects and ensure consistency from batch to batch, but you shouldn't feel obligated to go out and buy one. You won't need one to accomplish any of the recipes in this book.

How do you create such a library? Well, if you wanted, you could grow a new SCOBY from the liquid that your SCOBY came packaged in. It's so full of the organisms that give life to kombucha that it could produce a new SCOBY where there was none before. But growing a SCOBY this way can be painstakingly slow, so if you want to produce new SCOBYs for different-flavored kombuchas, it's best to grow from cuttings—literally cutting off a piece of a SCOBY and transferring it to its own syrup bath. Alternatively, you can keep one large SCOBY active in a neutral base of plain sugar syrup and take cuttings from it each time you start a new batch of kombucha.

(We advise against trying to grow a SCOBY from store-bought bottles of kombucha, because even varieties that advertise live cultures are sealed shut, starving the SCOBY of oxygen. Without knowing how long a bottle has been sitting on the shelf, there's no guarantee that the microbes are healthy enough to produce a new SCOBY.)

Finally, you should always be prepared for the very real possibility that your SCOBY might die. There are many bases you might think would be delicious as a kombucha but may, in fact, harbor natural antifungal or antibacterial agents that can kill a SCOBY. The first time we tried making kombucha from black-garlic stock, the kombucha took somewhere in the neighborhood of 20 days to acidify properly—more than twice as long as other varieties. We'd overlooked the garlic's natural chemical defense mechanisms. It contains a sulfur-based compound called allicin, which gives garlic its aroma and also fights off fungi.

We suspect that the allicin in the stock was interfering with the replication of the yeast in the SCOBY. Fortunately, however, some of the yeast took hold. By the time we started the following batch, we had a healthy SCOBY specialized in fermenting in the presence of allicin. Fermentation is evolution in real time, and it's a fascinating process in which to take part.

A refractometer determines the sugar content of a solution by measuring the refraction of light.

SYRUP

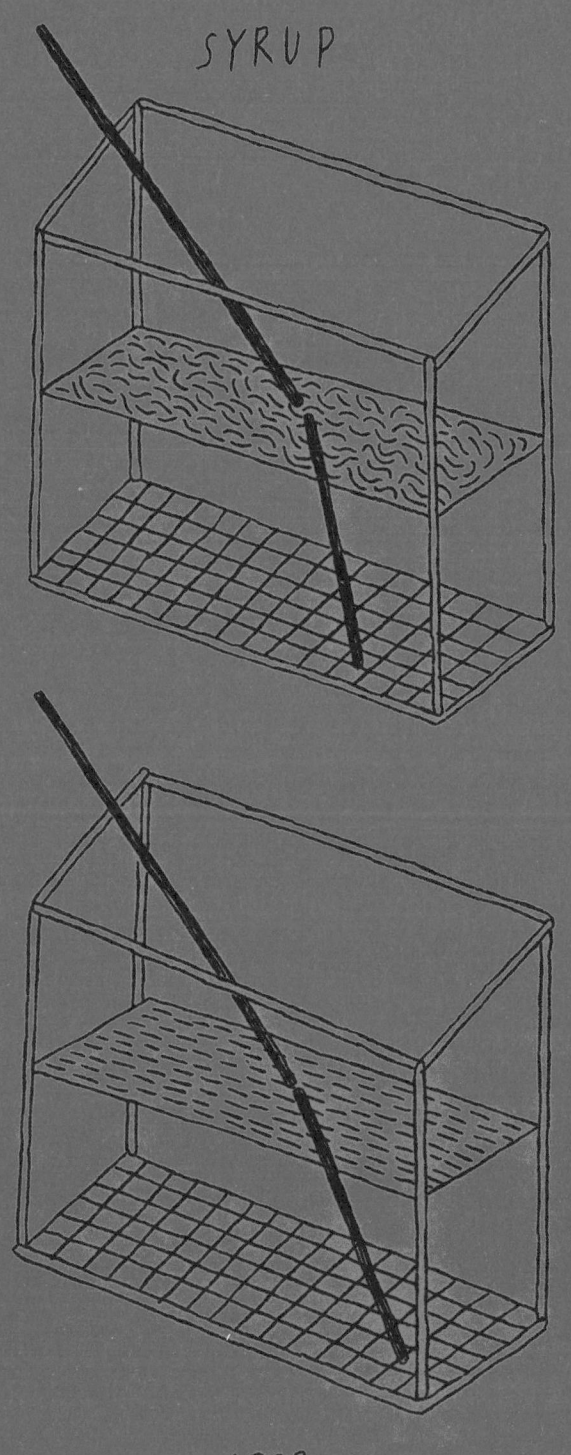

WATER

Above: A properly stored SCOBY can be used to ferment multiple batches of lemon verbena kombucha.
Opposite: Fresh lemon verbena has an electrifying aroma capable of filling a room.

Lemon Verbena Kombucha

Makes 2 liters

240 grams sugar

1.76 kilograms water

20 grams dried lemon verbena

200 grams unpasteurized kombucha
(or the liquid that comes with
a packaged SCOBY)

1 SCOBY (see Sources, page 448)

We'll begin our kombucha making with a recipe that closely follows the method for brewing a classic kombucha made from sweetened tea. The only difference is that our base liquid is an herbal infusion rather than a tea infusion. You can easily substitute a different herb or tea in place of lemon verbena, if you like; you'll still get a good handle on how to manipulate the various factors that affect the flavor, sweetness, acidity, and fermentation time of your kombucha.

The geology in Copenhagen leads to hard, mineral-rich tap water, which can distort the flavor of the kombucha. We filter the water for our ferments with a reverse-osmosis system. If you live somewhere with softer water, there's no reason tap water would harm your ferment, but if in doubt, filter your water.

Equipment Notes

Kombucha doesn't need much equipment other than a glass or plastic container of at least 2.5-liter capacity. Don't use metal containers—they can react negatively with the acid in the kombucha; plus, you won't be able to see what's going on inside. A SCOBY needs access to oxygen, so avoid vessels with tapered necks, like carboys. Large, wide-mouth canning jars work fantastically; clear plastic buckets and tall Tupperware also do the job nicely. You'll also need cheesecloth or a breathable kitchen towel to cover the vessel, and larger rubber bands to

Lemon Verbena Kombucha, day 1

Day 4

Day 7

secure it. And as with any of the sensitive microbes in this book, the SCOBY is best handled while wearing nitrile or latex gloves.

In-Depth Instructions

To begin, dissolve the sugar in a small amount of the water. (You only need a 1:1 ratio of water to sugar to completely dissolve the sugar, so heating the full amount of water is a waste of time. Plus, you'll have to wait for the water to cool before you introduce your SCOBY—yeast and acetic acid bacteria can't survive temperatures warmer than 60°C/140°F.) Bring the sugar and 240 grams of the water to a boil in a medium pot. Remove the pot from the heat, add the lemon verbena, and allow it to steep, uncovered, for about 10 minutes.

Once the tea has steeped, stir in the remaining 1.52 kilograms water and strain the infusion through a fine-mesh sieve or chinois into the clean fermentation vessel.

To jump-start fermentation and to help prevent unwanted microbes from getting a foothold, backslop the infusion by adding the 200 grams unpasteurized kombucha to your vessel (which is 10 percent of the weight of your other ingredients). Ideally, you'll be backslopping with a previous batch of lemon verbena kombucha, or a complementary flavor. If this is your first batch, use the liquid that your SCOBY came packaged in. Stir well with a clean spoon.

Put on your gloves and carefully place the SCOBY into the liquid. It should float, but don't worry too much if it sinks— it sometimes takes a day or two to rise to the surface.

Cover the top of the fermentation vessel with cheesecloth or a breathable kitchen towel and secure it with a rubber band. Fruit flies *love* the scent of acetic acid and alcohol, and will be particularly drawn to your new kombucha, so you'll want to do everything you can to keep them out.

Label the kombucha with its variety and the start date so you can easily keep track of its progress.

SCOBYs work best in slightly warm settings. If you're brewing in the summertime, you'll probably notice that your kombucha finishes faster than in the winter. In Noma's fermentation lab, we keep our kombucha room at a steady 28°C/82°F to encourage speedy production, but you don't need to dedicate a whole room of your house to kombucha. It will ferment just fine, albeit slightly more slowly, at room temperature. If you like, you can place your kombucha close to a radiator or on a high shelf in the kitchen to provide an environment that's slightly warmer than room temperature.

As the days go by, you'll notice the SCOBY growing significantly, fueled by the sugar in the liquid. Every other day or so, peel back the cloth covering enough to get a good look at the SCOBY. It should extend out toward the sides of your vessel, while also thickening in the middle. You may also see it puffing up in some areas as the yeast releases carbon dioxide. If you notice the top of the SCOBY drying out, use a ladle to pour a little liquid over it. The liquid keeps the SCOBY acidified, staving off mold growth.

There are a few different ways to measure the progress of the kombucha itself. The simplest method is one you're already well equipped for: Taste it. At Noma, we look for our kombuchas to maintain the essence of their base ingredient, while developing complexity and a harmonious opposition of sweetness and acidity. Put more simply: It's done when it tastes good. The kombuchas we brew at the restaurant usually take 7 to 9 days to ferment to our desired taste. If you enjoy sour kombucha, then let it ferment for an extra day or two.

In the fermentation lab, we use equipment to measure the acidity and sweetness of our kombuchas in order to maintain consistency from batch to batch. A refractometer allows you to track sugar levels in the brew. Taking a measurement in the beginning lets you know how much sugar you started with, and each subsequent measure tells you how much is left. A pH meter or pH strips gauge acid content. Infused lemon verbena syrup will begin with a pH of just under 7, which is close to neutral. Backslopping with a previous batch of kombucha

125

Bottling Kombucha

Bottling kombucha will extend its shelf life and encourage carbonation. A day or two *before* you're happy with the flavor (gauging this point will come with experience), strain the liquid, transfer it to sterilized swing-top bottles (or regular beer bottles, if you have a capping tool), and move them to the refrigerator. The residual bacteria and yeasts in the liquid will continue to work, even in the fridge. Bottling traps the gases from fermentation, some of which will dissolve into the liquid. A kombucha fermenting in open air will have a slight effervescence, but bottling will increase the bubbliness.

Take care not to bottle your kombucha too early. If there's too much residual sugar in the kombucha, it will fuel an excess amount of carbon dioxide production, which can result in exploding glass bottles. To mitigate this risk, make sure your kombucha is close to where you want the finished product before bottling—around 8°Bx, if you're measuring with a refractometer. Be sure to keep bottles in the fridge and consume them within a couple of weeks.

should drop the pH to about 5. Fermentation further increases the acidity to between 4 and 3.5. If you're equipped and inclined, keep track of your kombucha's progress and measure the pH and sugar content of the final product so it's easier to replicate.

If colorful (pink, green, or black) mold shows up on your SCOBY, it means your base liquid probably wasn't acidulated enough at the outset. (Though a healthy SCOBY may develop slight variations in color.) Don't try to salvage the liquid or SCOBY in this instance, as pathogenic molds can produce harmful toxins that dissolve into the liquid. Trying to identify whether an invasive mold is malignant or benign isn't worth the risk. You can always brew more kombucha.

Once you're satisfied with your kombucha's flavor, put on a pair of gloves and remove the SCOBY. Transfer it to a plastic or glass container into which the SCOBY fits snugly, and cover with three to four times its volume in kombucha. Cover the container with cheesecloth or a breathable kitchen towel, and secure it with rubber bands. It's fine to let the SCOBY hang out at room temperature if you intend to make another batch within the next few days. If you're not using the SCOBY again soon, store it in the fridge until you're ready. (See "Caring for Your SCOBY," page 116, for more information.)

Strain the remaining kombucha through a sieve lined with cheesecloth or a fine chinois. Now you can enjoy it straight away, or save it for later consumption or use in a recipe. Kombucha will keep in the fridge in a sealed container for 4 to 5 days without much change in flavor. You can also freeze it in an airtight plastic container or vacuum-sealed bag if you've made a larger batch than you can use immediately. To freeze your kombucha, chill it in the fridge for a few hours to slow fermentation before packing it into the container or bag, or it could inflate and even burst before freezing solid.

It may take you a couple of tries to nail a kombucha you're happy enough with to take to work or school. That's fine! You can still use overfermented kombucha for syrups. Meanwhile, your SCOBY will happily dive into a new batch, so keep trying.

Transfer kombucha to swing-top bottles a couple of days before it's reached your desired acidity. It will continue to ferment and carbonate in the bottle.

1. Water, SCOBY, lemon verbena, sugar, and finished kombucha.

2. Make a syrup using the sugar and an equal weight of water.

3. Combine the syrup and lemon verbena and allow to steep before adding the remaining water.

4. Strain the infusion through a fine-mesh sieve into the clean fermentation vessel.

5. Backslop with the unpasteurized kombucha.

6. Place the SCOBY into the fermentation vessel and cover.

7. Measure the sugar content using a refractometer (optional) and check again after 7 days.

8. Use pH strips to check the acidity of the kombucha. When the pH has reached 3.5 to 4, the kombucha should be close to ready.

9. Remove the SCOBY. Strain and bottle the kombucha.

Combine kombucha syrup with a mild oil for a quick vinaigrette.

Suggested Uses

Kombucha Syrup

Almost any kombucha can be cooked down to a remarkable and complex syrup, but it works especially well with kombuchas that are approaching the point of being too sour. Pour about 450 milliliters of kombucha into a medium saucepan on the stove over medium-low heat. Let the liquid slowly evaporate until it's about one-quarter of its original volume and can coat the back of a spoon. The slower the kombucha reduces, the better—don't let it come to a boil or you'll cook out all the flavor.

The next time you cook a batch of pancakes, drizzle a bit of this syrup over the top. It won't be as sweet as maple syrup, so if you have a sweet tooth, you may want to add a dusting of powdered sugar. As a dessert, spoon some lemon verbena kombucha syrup over good-quality ice cream. Bonus points if you sprinkle freshly picked lemon verbena on top.

Lemon Verbena Kombucha Vinaigrette

Blending together equal parts mild oil, such as virgin rapeseed or avocado, and lemon verbena kombucha syrup will produce a thick vinaigrette—sweet, sour, and creamy. You'll need to taste for salt and acidity and adjust accordingly, but the result is an exceptional dressing for root vegetables. For a first-rate side dish, toss salt-baked beets in the vinaigrette and garnish with torn fresh basil leaves and chopped pistachios.

131

Backslopping with a previous batch of kombucha simultaneously lowers the pH and adds a healthy starting population of bacteria and yeasts.

Rose Kombucha

Makes 2 liters

240 grams sugar
1.76 kilograms water
200 grams wild rose petals
200 grams unpasteurized kombucha
    (or the liquid that comes with
    a packaged SCOBY)
1 SCOBY (see Sources, page 448)

In Denmark, you'll find wild roses growing everywhere. Their loose blossoms of small petals aren't as visually impressive as cultivated hybrid roses, but their perfume and flavor are stunning. This kombucha will retain the sweet rose fragrance and be balanced by the bright acidity of the fermentation. If you can't get your hands on wild roses, choose flowers with a pronounced scent that have not been sprayed with chemicals or contaminated in any way; petal size doesn't matter.

The in-depth instructions for Lemon Verbena Kombucha (page 123) serve as a template for all the kombucha recipes in this chapter. We recommend you read that recipe before starting in on this one.

Bring the sugar and 240 grams of the water to a boil in a medium pot, stirring to dissolve the sugar. Remove from the heat and add the remaining 1.52 kilograms water to cool it down quickly.

Once the sugar syrup cools to room temperature, transfer it to a blender along with the rose petals and blend. You won't get a perfect puree, but the petals should be reduced to small bits. Transfer the liquid to a container, cover, and refrigerate overnight to infuse.

Rose Kombucha, day 1

Day 4

Day 7

The following day, strain the rose syrup through a fine-mesh sieve into the fermentation vessel. Backslop the infusion by stirring in the 200 grams unpasteurized kombucha. Wearing gloves, carefully place the SCOBY into the liquid. Cover the top of the fermentation vessel with cheesecloth or a breathable kitchen towel and secure it with a rubber band. Label the kombucha and set it in a warm place.

Leave the kombucha to ferment, tracking its progress each day. Make sure the top of the SCOBY doesn't dry out; use a ladle to moisten it with some of the liquid, if necessary. Once you're happy with the flavor of your kombucha—probably between 7 and 10 days from the start—transfer the SCOBY to a container for storage and strain the kombucha. Consume immediately or refrigerate, freeze, or bottle it.

Suggested Uses

Rose-Plum Sauce for Duck

Rose kombucha can serve as the base for a tart, floral plum sauce for roasted duck breast or grilled halloumi. Blend roughly equal amounts of lacto plum flesh (page 69) and rose kombucha until smooth—200 grams of each should make plenty of sauce. (If you don't have any lacto plums on hand, you can make do with half that amount of store-bought umeboshi plums.) Pass the mixture through a fine-mesh sieve and serve in small ramekins with a drizzle of olive oil and a sprinkle of cracked Sichuan peppercorns.

Gin and Rose Cocktail

At Noma, the juice pairings menu (where a lot of our kombuchas feature) is meant as an alternative to wine pairings, but that's not to say you can't mix kombucha with alcohol. This Saturday night, try muddling a handful of fresh berries into 50 milliliters rose kombucha, and combine with 28 milliliters gin (or vodka) before straining it into a rocks glass over ice.

Berry-Rose Coulis

Blend together 500 grams rose kombucha and 250 grams of the season's best berries. The longer you blend, the more pectin you'll release from the berries, and the thicker the puree will be. At this point, you could serve it as a lovely summer refreshment—no straining necessary—or you could pass the mixture through a fine-mesh sieve, turning it into a vibrant coulis for macerating more fresh fruit or topping ice cream or panna cotta.

Passing your puree through a tamis creates a luxurious texture.

Apple kombucha is one of the simplest yet most
versatile kombuchas in this book.

Apple Kombucha

Makes 2 liters

2 kilograms unfiltered apple juice
200 grams unpasteurized kombucha
    (or the liquid that comes with a
    packaged SCOBY)
1 SCOBY (see Sources, page 448)

Juicing your own apples will allow you to use local varieties and create a blend to your liking, but feel free to use a good-quality store-bought unfiltered apple cider; farmstands often sell fresh-pressed cider in season. Because the juice is naturally sweet, you won't need to add sugar to this recipe.

The in-depth instructions for Lemon Verbena Kombucha (page 123) serve as a template for all the kombucha recipes in this chapter. We recommend you read that recipe before starting in on this one.

Pour the apple juice into the fermentation vessel. Backslop by stirring in the 200 grams unpasteurized kombucha. Wearing gloves, carefully place the SCOBY into the liquid. Cover the top of the fermentation vessel with cheesecloth or a breathable kitchen towel and secure it with a rubber band. Label the kombucha and set it in a warm place.

Leave the kombucha to ferment, tracking its progress each day. Make sure the top of the SCOBY doesn't dry out; use a ladle to moisten it with some of the liquid, if necessary. Once you're happy with the flavor of your kombucha—probably between 7 and 10 days from the start—transfer the SCOBY to a container for storage and strain the kombucha. Consume immediately or refrigerate, freeze, or bottle it.

137

Apple Kombucha, day 1

Day 4

Day 7

Suggested Uses

Apple Kombucha Herb Tonic

Blending apple kombucha with fresh herbs infuses the liquid with ethereal aromatic qualities. In Copenhagen, we're fortunate to be able to take a walk around the neighborhood and find young Douglas fir branches to make a brisk apple-pine tonic. (Whir 25 grams fresh fir needles with 500 grams apple kombucha in a blender, strain, and serve.) But you can also find plenty of suitable dance partners for apple kombucha at your local market. Use a stand blender to whir half a bunch of basil or 10 grams picked rosemary needles with 500 grams apple kombucha. Strain through a fine-mesh sieve for an invigorating pick-me-up.

Apple-Vegetable Smoothie

Blending cooked vegetables with fruit kombuchas is an absolutely delicious way to get a little fiber (and also a great way to sneak more vegetables into your kids' diets). Good matches for apple kombucha include spinach, sorrel, cabbage, or baked beets (which also pair well with rose kombucha). Because the vegetables are so full of fiber, they will thicken up in a blender nicely. Aim for a 4:1 ratio of kombucha to vegetable, and blend for at least a minute before passing it through a fine-mesh sieve and serving.

Blending apple kombucha with herbs
(in this case, Douglas fir pine needles) makes
a bright, refreshing tonic.

Elderflower kombucha is like the bottled flavor
of Scandinavian summers.

140

Elderflower Kombucha

Makes 2 liters

240 grams sugar
1.76 kilograms water
300 grams fresh elderflower blossoms
200 grams unpasteurized kombucha
    (or the liquid that comes with
    a packaged SCOBY)
1 SCOBY (see Sources, page 448)

Elderflower is one of the hallmark flavors of summer in Scandinavia. The tiny, sweetly fragrant white blossoms make their way onto Noma's menu year after year. You'll find them in early summer in many temperate climates across the Northern Hemisphere, as well as in parts of Australia and South America.

The in-depth instructions for Lemon Verbena Kombucha (page 123) serve as a template for all the kombucha recipes in this chapter. We recommend you read that recipe before starting in on this one.

Bring the sugar and water to a boil in a medium pot, stirring to dissolve the sugar. Meanwhile, put the blossoms in a nonreactive heatproof container. Pour the hot syrup over the elderflower blossoms, then allow to cool to room temperature. Once cool, cover the container and transfer to the fridge to infuse overnight.

The following day, strain the elderflower syrup through a fine-mesh sieve into the fermentation vessel, pressing on the blossoms to extract as much liquid as possible. Backslop the infusion by stirring in the 200 grams unpasteurized kombucha. Wearing gloves, carefully place the SCOBY into the liquid. Cover the top of the fermentation vessel with cheesecloth or a breathable kitchen towel and secure with a rubber band. Label the kombucha and set it in a warm place.

Elderflower Kombucha, day 1

Day 4

Day 7

Leave the kombucha to ferment, tracking its progress each day. Make sure the top of the SCOBY doesn't dry out; use a ladle to moisten it with some of the liquid, if necessary. Once you're happy with the flavor of your kombucha—probably between 7 and 10 days from the start—transfer the SCOBY to a container for storage and strain the kombucha. Consume immediately or refrigerate, freeze, or bottle it.

Suggested Use

Elderflower Crème Fraîche

This experimental mix of kombucha and dairy has shown up here and there on our menu at Noma, and you might find a bevy of uses for it at home, too. Mix together 800 grams cream, 200 grams whole milk, and 200 grams elderflower kombucha. Cover with cheesecloth and leave to ferment at room temperature for 2 to 3 days. The cream will thicken and the flavor of the elderflower will impart a floral accent to the soured cream, like a bloomy soft-rind cheese.

Try mixing a bowl of blanched and shucked fresh sweet peas with a good spoonful of this cream as a bright and refreshing start to an outdoor lunch. Garnish with slivers of crunchy radishes and a smattering of freshly picked herbs, such as lemon thyme, lemon verbena, or chervil.

Combine elderflower kombucha, cream, and milk
to make a distinctly floral crème fraîche.

143

Coffee kombucha is a fantastic way to get more life
out of used coffee grounds.

Coffee Kombucha

Makes 2 liters

240 grams sugar

1.76 kilograms water

730 grams leftover coffee grounds,
    or 200 grams freshly ground coffee

200 grams unpasteurized kombucha
    (or the liquid that comes with
    a packaged SCOBY)

1 SCOBY (see Sources, page 448)

Coffee kombucha offers a second life to used coffee grounds, which still have plenty of flavor to give up. If you prefer, you can use new grounds, but note that you'll use much less of the fresh stuff. Look for coffee that hasn't been roasted too dark, which can turn it quite bitter—a lighter roast will allow the complex fruitiness of a good coffee to shine through.

The in-depth instructions for Lemon Verbena Kombucha (page 123) serve as a template for all the kombucha recipes in this chapter. We recommend you read that recipe before starting in on this one.

Bring the sugar and 240 grams of the water to a boil in a medium pot, stirring to dissolve the sugar. Meanwhile, put the coffee grounds in a nonreactive heatproof container. Pour the hot syrup over the coffee, then add the remaining 1.52 kilograms water. Let the coffee mixture cool to room temperature, cover, and transfer to the fridge to infuse overnight.

The following day, strain the coffee liquid through a fine-mesh sieve lined with cheesecloth into the fermentation vessel. Backslop the infusion by stirring in the 200 grams unpasteurized kombucha. Wearing gloves, carefully place the SCOBY into the liquid. Cover the top of the fermentation vessel with cheesecloth or a breathable kitchen towel and secure it with a rubber band. Label the kombucha and set it in a warm place.

145

Coffee Kombucha, day 1

Day 4

Day 7

Leave the kombucha to ferment, tracking its progress each day. Make sure the top of the SCOBY doesn't dry out; use a ladle to moisten it with some of the liquid, if necessary. Once you're happy with the flavor of your kombucha—probably between 7 and 10 days from the start—transfer the SCOBY to a container for storage and strain the kombucha. Consume immediately or refrigerate, freeze, or bottle it.

Suggested Uses

Coffee-Kombucha Tiramisu

The next time you're having a dinner party, make a tiramisu, using coffee kombucha in place of coffee to soak your ladyfingers. Tiramisu is quite rich and sweet with custard, and the pleasantly vibrant bite of coffee kombucha acts as a perfect counterpoint.

Parsnips Glazed with Coffee Kombucha

Let's say you've got a pan of peeled and quartered parsnips, caramelizing gently in foaming butter on the stove. Two minutes before removing them from the pan, throw in a sprig each of sage and thyme, turn up the heat a bit, and deglaze with about 120 milliliters coffee kombucha. Swirl the pan, paying attention as the mixture thickens and begins to stick to the parsnips. At the last minute, add a big spoonful of butter and allow it to melt and glaze the parsnips. Remove from the pan and finish with a sprinkling of smoked salt.

Pan-roast parsnips and glaze them in coffee kombucha and butter.

Reducing maple kombucha to a (better) syrup brings
things full circle.

148

Maple Kombucha

Makes 2 liters

360 grams pure maple syrup
1.64 kilograms water
200 grams unpasteurized kombucha
    (or the liquid that comes with
    a packaged SCOBY)
1 SCOBY (see Sources, page 448)

Use a good-quality pure maple syrup—not the food-colored corn syrup you find at many grocery stores. The quality of your finished kombucha will only be as good as the ingredients that go into it.

The in-depth instructions for Lemon Verbena Kombucha (page 123) serve as a template for all the kombucha recipes in this chapter. We recommend you read that recipe before starting in on this one.

The sugar in maple syrup is already dissolved, so there's no need to apply heat, but you will have to add the water to dilute it to a sweetness of about 12°Bx. Pour the maple syrup, water, and the 200 grams unpasteurized kombucha into your fermentation vessel and stir. Wearing gloves, carefully place the SCOBY into the liquid. Cover the top of the fermentation vessel with cheesecloth or a breathable kitchen towel and secure it with a rubber band. Label the kombucha and set it in a warm place.

Leave the kombucha to ferment, tracking its progress each day. Make sure the top of the SCOBY doesn't dry out; use a ladle to moisten it with some of the liquid, if necessary. Once you're happy with the flavor of your kombucha—probably between 7 and 10 days from the start—transfer the SCOBY to a container for storage and strain the kombucha. Consume immediately or refrigerate, freeze, or bottle it.

149

Maple Kombucha, day 1

Day 4

Day 7

Suggested Uses

Quatre Épices Cocktail

For a Christmas-y beverage, cold-steep 25 grams quatre épices in 500 grams maple kombucha for a few days. To make your own four-spice blend, toast 2 parts white pepper, 1 part cloves, 1 part grated nutmeg, and 1 part ground ginger in a dry pan before adding them directly to the kombucha. Keep covered in the fridge for at least 2 days and strain before serving. Once the kids have been put to bed, add a splash of coffee liqueur to make things even more festive.

Maple Kombucha Syrup

Bring maple kombucha full circle by reducing it back into a fantastically sweet and sour syrup. Heat 1 liter maple kombucha in a saucepan over low heat and reduce it slowly until it can coat the back of a spoon. Don't be tempted to speed things up—you'll lose much of the aroma and nuance if you boil it. Once it's reduced, allow the syrup to cool to room temperature before storing it in an airtight container in the fridge. Maple kombucha syrup is incredible with chocolate. You can see for yourself by making your favorite chocolate mousse and glazing it with this syrup.

Kombucha BBQ Sauce

A great use for maple kombucha syrup (or any other kombucha syrup, for that matter) is in a classic barbecue sauce. While many barbecue sauce recipes call for acidity in the form of apple cider vinegar, they still tend to lean toward the sweet side of things. By replacing the sugar with kombucha syrup, you still get the perception of sweetness without the sugar, plus more acidic twang to cut through fattier items like ribs or chicken legs.

Cold-infuse maple kombucha
with a classic blend of quatre épices
for a nonalcoholic cocktail. Add
a splash of coffee liqueur to make it
an alcoholic one.

151

Made from fruit puree, mango kombucha
has more body and texture than other kombuchas.
This recipe uses Kent mangoes, but there are
dozens of different cultivars that will all ferment
with unique flavors.

Mango Kombucha

Makes 2 liters

170 grams sugar

970 grams water

800 grams diced peeled ripe mango
 flesh

200 grams unpasteurized kombucha
 (or the liquid that comes with
 a packaged SCOBY)

1 SCOBY (see Sources, page 448)

All the other kombuchas in this book are thin liquids with a viscosity not so different from water. While developing dishes for our pop-up restaurant in Tulum, Mexico, we wanted to add some texture to the kombuchas in our juice pairings—something pleasing yet fluid enough to ferment properly. If the purees were too thick, the microbes would get jammed up, unable to move around. We found that blending equal parts mango flesh and water for a minute produced exactly what we were looking for.

The opacity of the mango puree means a refractometer won't be able to measure the sugar content, so you'll really need to use your taste buds to inform your decisions. Generally speaking, the sweeter and riper the mango, the better.

The in-depth instructions for Lemon Verbena Kombucha (page 123) serve as a template for all the kombucha recipes in this chapter. We recommend you read that recipe before starting in on this one.

Bring the sugar and 170 grams of the water to a boil in a medium pot, stirring to dissolve the sugar. Remove from the heat and let cool to room temperature.

Put the mango and remaining water in a blender and process until you have a smooth puree, about 1 minute. You may need to work in batches to fit your blender size.

Mango Kombucha, day 1

Day 4

Day 7

Strain the mango puree through a fine-mesh sieve into the fermentation vessel, then stir in the simple syrup and the 200 grams unpasteurized kombucha. Wearing gloves, carefully place the SCOBY into the liquid. Cover the top of the fermentation vessel with cheesecloth or a breathable kitchen towel and secure it with a rubber band. Label the kombucha and set it in a warm place.

Leave the kombucha to ferment, tracking its progress each day. The mango puree will likely separate into a thin liquid with a thicker layer floating above it, which is fine, but it does mean that the SCOBY might have a harder time getting the oxygen it needs. To assist your microbial mother, slide a clean spoon beneath the SCOBY and stir the puree together each day, being careful not to disturb the SCOBY too much. You can also ladle some liquid over the SCOBY to wash off the heavier mango solids.

Once you're happy with the flavor of your kombucha—probably between 7 and 10 days from the start—transfer the SCOBY to a container for storage, scraping off as much of the mango puree as possible. Strain the kombucha through a fine-mesh sieve lined with cheesecloth. Consume immediately or refrigerate, freeze, or bottle it.

Suggested Uses

Mango Gazpacho

Kombuchas with more body and texture do well as the foundation of a cold soup. For a different spin on gazpacho, chop, squeeze, and strain the juice from 3 beefsteak tomatoes. Combine the juice with 500 milliliters mango kombucha and season with salt. Add a mixture of cooked vegetables—sliced grilled asparagus, diced celery, fava beans, or sliced grilled scallions come to mind—and a spoonful of pomegranate seeds for crunch. Finish it off with grilled parsley and cilantro leaves.

Mango-Lemongrass Vinaigrette

Kombuchas made from purees are a little less versatile than those made with tisanes or other infusions, but you can still make good use of the extra viscosity and sweetness of a mango kombucha. Muddle a stalk of lemongrass and a fistful of cilantro in 500 milliliters mango kombucha, and let it sit for 15 minutes before straining. Add a healthy drizzle of spicy chili oil and season with a pinch of sea salt. Use the vinaigrette for whole roasted fish or grilled vegetables like bok choy or avocados. Or use the same sauce to glaze a roast pork shoulder or scallops on the grill.

Muddle fresh herbs with mango kombucha to create a beverage (or cold soup) with even more character.

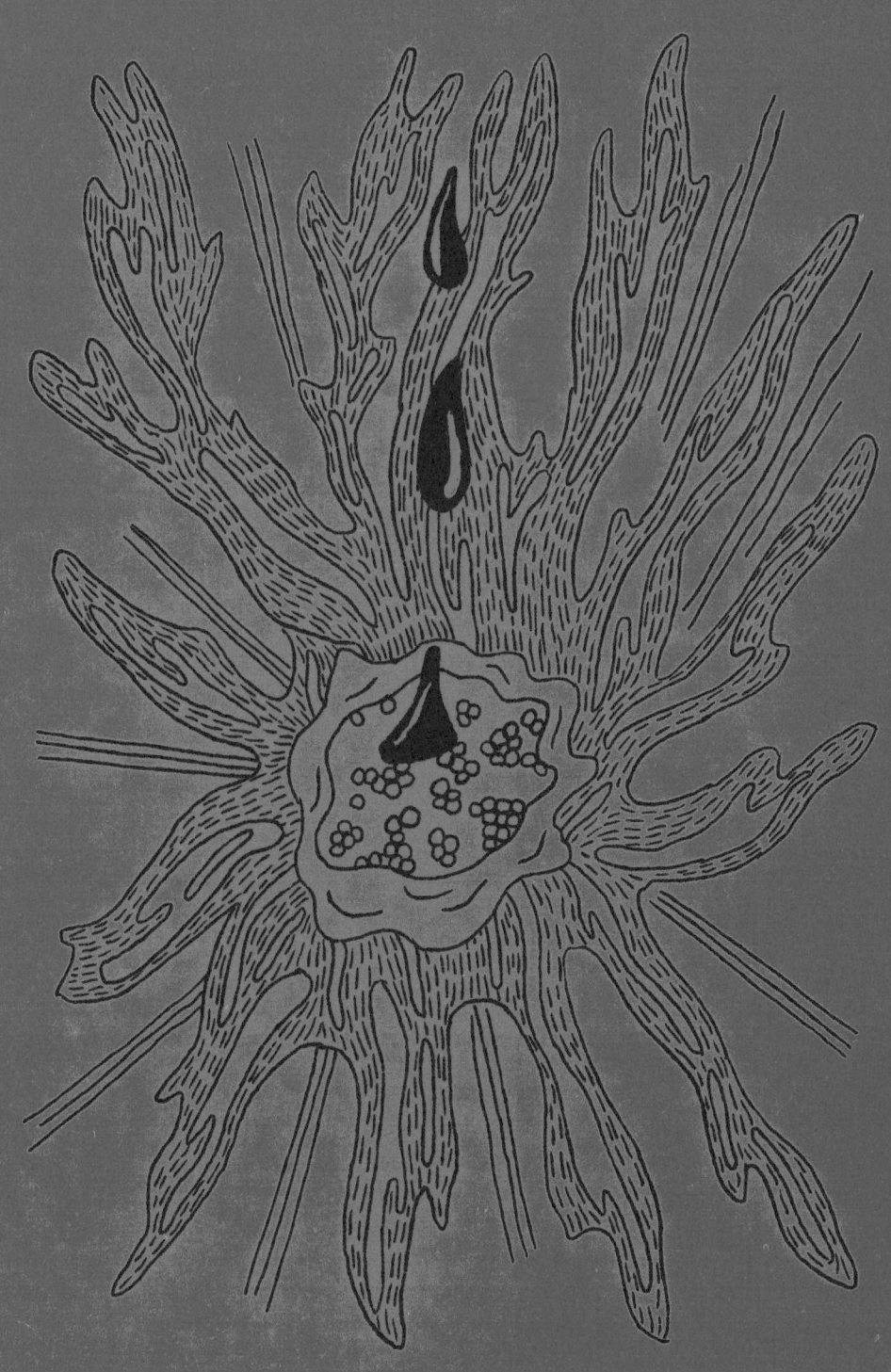

Vinegar

—

Perry Vinegar 173

Plum Vinegar 183

Celery Vinegar 187

Butternut Squash Vinegar 191

Whiskey Vinegar 195

Gammel Dansk Vinegar 198

Elderberry Wine Balsamic 201

Black Garlic Balsamic 206

Vinegar Improves Practically Anything

Throughout Europe and probably the entire Western world, adding a splash of vinegar is the prevailing method of injecting a bit of freshness into whatever you're cooking. The next time you have some store-bought orange marmalade, add a little vinegar and a pinch of salt to make it instantly more vibrant. If you're making your own ice cream—depending on the type you're making—a little fruity vinegar will give it an unexpected edge. And very few cooked vegetables or fruits aren't improved with a hit of vinegar.

When Noma first opened, vinegar was more or less the only tool we could depend on to instill our food with acidity. We'd pair the likes of, say, beets and apples, and find that they needed a bridge—something fruity, with acidity to link the earthiness and sweetness of the two main ingredients. Aged apple vinegars were often up to the task.

Before we really found our way with lacto-fermentation, most of the pickling we did at Noma was with vinegar. Here in Scandinavia, vinegar pickles are everywhere, likely because they're so simple to make: Combine one part water, one part vinegar, and a little salt and sugar; add your fruits or vegetables; and let them sit. Vinegar-pickling plays less of a role at Noma now, but we still do a little bit of it with ingredients like shoots, mushrooms, and seasonal flowers. Potent blooms like elder-flower, rose petals, colt's foot, chamomile, or dandelion flowers are left to mature in apple vinegar for at least a few weeks in the fridge, before the pickled flowers find their way into all sorts of dishes, from roasted bone marrow to desserts. As a happy side effect, the vinegar takes on the flowers' hue and fragrance and can be used to bring tartness to both sweet and savory dishes, long after the pickles themselves are gone. The same method can be applied to nice effect with fresh fruits. Many of the fruit vinegars you find at grocery stores are produced by soaking fruit in neutral-flavored vinegars.

Balanced acidity is crucial to a meal at Noma, which is why we've always found vinegar to be such a powerful ingredient. The word *vinegar* comes from the Latin *vinum acer*—literally "sour wine." But of course, that only scratches the surface of

Berries and Greens Soaked in Vinegar for One Year,
Noma, 2016

This palate-cleansing course includes lacto-fermented
red gooseberries and wild cherries served with
chanterelles marinated in pumpkin vinegar, wild rose
petals in rosehip vinegar, elderflower and colt's
foot in apple vinegar, and black currant shoots in
spruce vinegar.

159

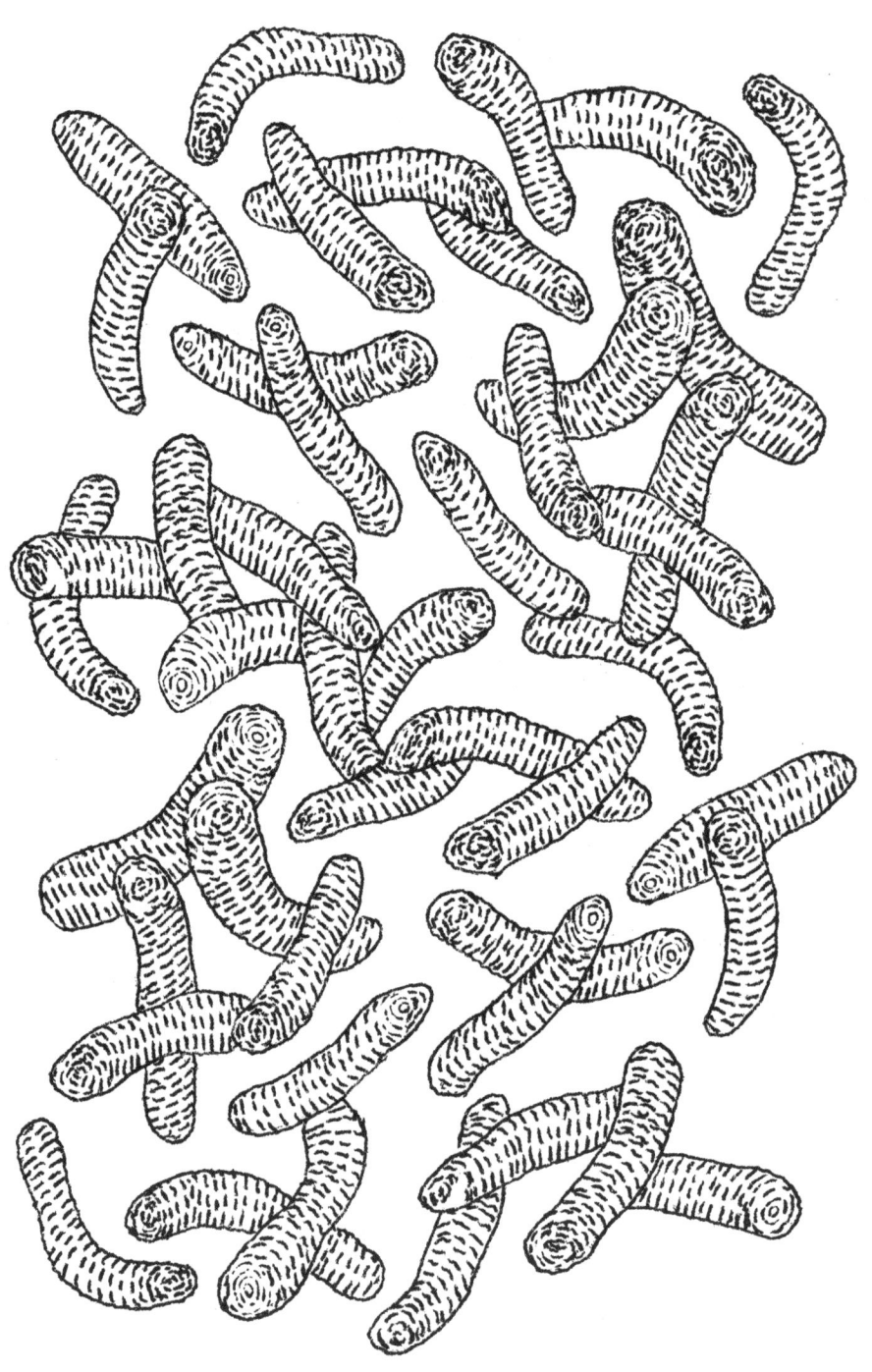

160

what vinegars can be. There are aged vinegars, such as balsamic, that have texture and sweetness. There are very strong vinegars that cut through anything with their acidity. On the other side of the spectrum, there are vinegars with very little acidity (only 1% or 2%) that you can drink straight from the bottle, or use as sauces in their own right. We make a vinegar from leftover fennel tops in the summer that's a perfect example of the latter. The lower acidity lets the original flavor shine through, bringing an additional layer of brightness without detracting from the fennel.

You'll find dozens of varieties of vinegar in a well-equipped supermarket, so even if you don't choose to make your own vinegar, there's no reason you can't begin experimenting with the different applications we suggest in this chapter. But if you're ready to go further down the rabbit hole, read on.

The Acid Test of Time

Vinegar is a kitchen pillar that is so ubiquitous and familiar that many people don't think of it as a product of fermentation.

In fact, vinegar is made by the fermentation of alcohol into acetic acid by a large family of obligate aerobic bacteria (bacteria that need air to function). These acetic acid bacteria (AAB) encompass a wide array of species. They are omnipresent and airborne and found on surfaces of most living things, including you.

As with kombucha, vinegar is a product of the collaboration between yeast and bacteria. First, yeast converts sugar to alcohol, then AAB convert the alcohol to acetic acid. The difference is that vinegar-makers will often select yeasts with a certain maximum threshold for alcohol, meaning the yeast will die off before it's had the chance to consume all the sugars in the base liquid. (Alternatively, they will sometimes heat the alcohol to kill the yeast.) Otherwise, many yeasts aren't adapted to survive acetic acid and will end up dying once the AAB takes over. Thus, whereas kombucha will continually grow more and more acidic until all available sugar has been converted to alcohol (and subsequently acid), vinegar will plateau at a certain acidity.

161

Varieties of vinegars around the world are as diverse as the cultures that produce them, often reflecting the alcoholic spirits native to the region. East to West, we can find vinegars fermented from rice, sorghum, millet, barley, kiwi, apple, honey, berries, coconut, and beyond. The fermentable sugars in most of these products are readily available, allowing yeast to get to work straightaway. With grains like rice and barley, enzymes must first break down the starch in the grain into fermentable sugars. (You can read about this in more detail in the koji chapter, page 211.)

The earliest vinegars were derived from products that had already been fermented into alcohol, and they were also almost certainly an accident. Before the advent of microbiology, the reasons alcohol would sour into vinegar were a mystery. As surely as the sun would rise and set, so too would wine become vinegar if left in the open air. The cause was anyone's guess.

But that's not to say that people were unfamiliar with the process of fermentation. People have been making alcohol from fruit for the entire duration of human civilization. In Iran, shards of urns dating back to 6000 BCE—excavated from what was once the kitchen of a Neolithic abode near the Zagros Mountains—display yellowish-red stains from wine. Thousands of years later, the ancient Egyptians were producing alcoholic grape beverages of their own. There's evidence that as early as 3000 BCE, Egyptian kings were being buried with jars of wine in their tombs. Archaeologists have examined these jars and also found the residues of vinegar.

While ancient civilizations may not have known exactly why fruit turned to wine or wine to vinegar, they understood *how* it was happening, as evidenced by an Egyptian papyrus from the Ptolemaic period, *The Instruction of Ankhsheshonq*, which contains a note on the preservation of wine: "Wine matures as long as one does not open it." Two millennia later, the demystification of vinegar was much more than a culinary leap—it upended the way we understood nature.

Humans have been fermenting wine, and subsequently vinegar, since at least 6000 BCE.

Antoine Lavoisier: a pillar of modern chemistry and one of the first to understand how wine becomes vinegar.

Until midway through the eighteenth century, the prevailing wisdom was that everything on earth was composed from four basic elements: fire, water, earth, and air. Antoine Lavoisier, one of the founders of modern chemistry, was the first person to suggest that air was not a pure, immutable substance, but rather a combination of components, including oxygen (a term he coined from the Greek words for "acid former"). Through rigorous experimentation using nonmetals such as sulfur and phosphorous, he correctly deduced that oxygen was being removed from the surrounding air when the elements were burned. The products of those reactions were acids. By extrapolating these results, Lavoisier came to the conclusion that the transformation of wine into vinegar occurred because of oxygen in the air, through the process of "oxidation" as carried out by AAB.

This cognitive leap spread across Europe and led to advancements in vinegar production that exploited the idea of oxidation. Vinegar-makers were able to speed up the process by increasing the surface area of wine. German vinegar-makers developed "the quick process" to drip wine through loosely packed wood chips while simultaneously blowing the liquid with fresh air. Hundreds of years later, artisans still employ the method.

We take advantage of the same idea in our own way at Noma. Using a common air pump—the kind you'll find in the aquarium section of any pet store—we send air through our would-be vinegar, providing AAB with the oxygen they need to work quickly. By treating our bacteria as if they were pet goldfish, we can cut our fermentation time from a few months to a couple of weeks. You'll find more details in the in-depth recipe for Perry Vinegar (page 173).

The Quicker
Quick Process

At Noma, we make several vinegars the traditional two-stage way—fermenting alcohol from a raw product, then allowing AAB to produce vinegar from the alcohol—using our take on the quick method.

From start to finish, a two-step vinegar fermentation looks like this:

1. Inoculate sweet fruits or vegetables with yeast. Allow to ferment for 10 to 14 days, or until the liquid has an alcohol by volume (ABV) of 6% to 7%.

2. Strain the alcohol and heat to 70°C/158°F to kill any remaining yeast.

3. Transfer the liquid to large mason jars and backslop with a previous batch of vinegar. (See page 33 for more on backslopping.)

4. Run an air pump attached to an air stone (a piece of porous rock or metal that diffuses air in small bubbles). Ferment for 10 to 14 days, or until all the alcohol has been converted to acid.

That's how we produce excellent vinegars from pears, apples, and plums. However, you can also make compelling vinegars from products that can't be fermented into alcohol. Vegetables like celery or fennel contain too little sugar for yeast to produce enough alcohol for AAB to work with. Even if the yeast could convert all the available sugar to alcohol, it would take a long time, leaving the liquid vulnerable to infection by unwanted microbes, which could introduce off-flavors or spoil the batch outright.

In order for AAB to produce a vinegar of about 5% acidity, they need to work in a liquid with a total alcohol content of 6% to 8% ABV. Fruits or vegetables with a sweetness of less than 14°Bx (see page 118 for more about the Brix scale) will generally not have enough sugar to reach the required ABV

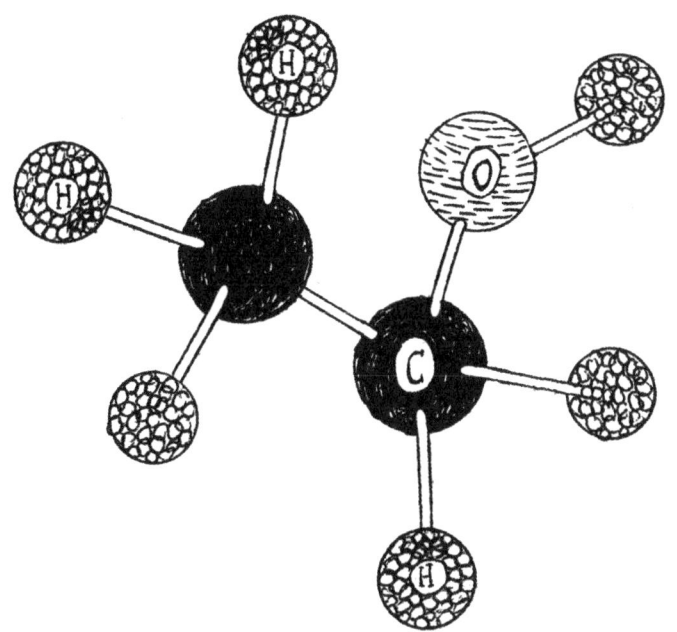

Through fermentation,
ethanol (C_2H_5OH) turns into . . .

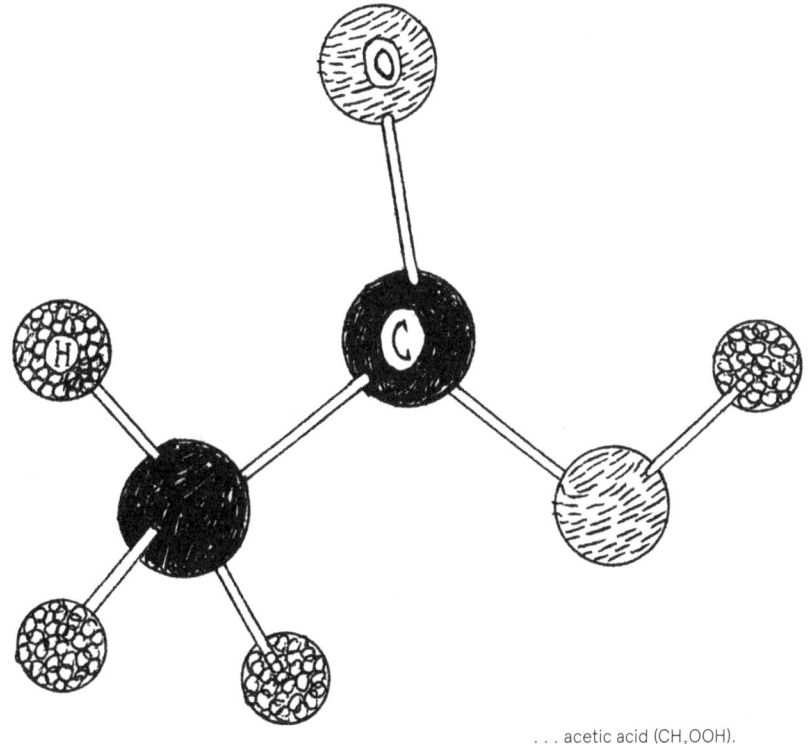

. . . acetic acid (CH_3OOH).

165

with enough left over to provide the sweetness you need for a balanced vinegar. In cases like this, we make up the difference by feeding the AAB with distilled ethanol.

Ethanol, or ethyl alcohol, is what's found in alcoholic beverages. When sold in its pure form, ethanol is sometimes referred to as NGS (neutral grain spirit) or "rectified spirit"—a distilled product with a maximum ABV of 96% (with the other 4% being water). There are a few brands of alcohol that are bottled near that percentage, such as Everclear and Gem Clear in North America, or Primaspirit in Europe, all of which work perfectly for vinegar production. Avoid "denatured ethanol" or anything with a listed ABV of 100%. Don't use any product with isopropyl alcohol, methyl ethyl ketone, or anything other than ethyl alcohol or water as an ingredient. Such products are not safe for consumption. If you can't find ethanol, another low-flavor alcohol like vodka will work, although you'll need more of the spirit by volume to yield the same starting alcohol content. For example, if a recipe calls for 100 grams of 96% ethanol, 130 grams of 75% (150-proof) vodka is required to reach the same alcohol content. (See "If You Can't Find Ethanol, Use Vodka," page 189.)

By adding ethanol to fruit or vegetable juice, vinegar making effectively becomes a one-stage fermentation. AAB can get to work right way, without ever needing yeast to propagate and create alcohol from the base. The sugar in the juice will go unfermented and provide balance to the final product. By going this route we've made vinegars from seaweed stocks, carrots, cauliflower, beets, squash, and more.

Perhaps an existential question may occur to you: "Why not just add acetic acid to the juice and skip the whole fermentation process?"

We strive for complex and intriguing flavors in our ferments at Noma, and as with many cooking processes, what you gain in a shortcut leaves much on the cutting-room floor. White vinegar is fermented from pure ethanol and then diluted to an acetic acid content of around 5%. It's relatively harsh and unnuanced, though it has its place in some cuisines. When AAB ferment vinegar, they can produce metabolites other than acetic acid, such as gluconic acid and ascorbic acid, that bring character and depth to vinegar. Plus, there are any number of unpredictable secondary reactions taking place in the fermentation process—some flavors get muted and new ones come forth. These are the qualities of good vinegar. It's the minutiae that make the difference.

Spirit Vinegars

Another question you may be asking is whether you can take fresh fruits and vegetables out of the vinegar equation altogether. Sure, why not? You can also transform distilled spirits into vinegars, as long as you dilute them first or burn off some of the alcohol—high-alcohol contents will hinder AAB.

Dilution is the most straightforward way of making vinegar from spirits, but you have to be careful not to water things down to the point of unrecognizability. By the time you've watered down something like a mild plum aquavit, you may as well be fermenting a flavorless vodka. On the other hand, consider Gammel Dansk, the classic Danish bitters, which is so heavily infused with botanicals that diluting it to 8% ABV will barely put a dent in its flavor.

In order to retain the character of more delicate spirits like bourbon or schnapps, it's better to remove the alcohol rather than dilute it. Flambé the spirit in a saucepan over high heat until the flames subside. You'll lose a lot of the liquid in the process, but what you're left with is a nearly alcohol-free version of your original spirit. The flavor will also be concentrated, so you'll want to add water to bring things back into balance. Measure the volume of the liquid, and add some of the original spirit to bring the ABV back up to the neighborhood of 8%. From there, proceed as you would with the second stage of any vinegar fermentation, by back-slopping with some unpasteurized vinegar, aerating the mixture, and waiting patiently.

167

The Angel's Share

Unlike similar fermentations such as kombucha, vinegar can receive extensive upgrades to its taste and texture through aging. Because AAB stop producing acid once all the available alcohol has been consumed, vinegar can sit for decades without growing more sour. Aging vinegar can instill layers of flavor by way of the aging vessel, and via evaporation and slow Maillard reactions that occur over the course of months and years. (See "Really Slow Cooking," page 405, for more about Maillard reactions.)

The most famous aged vinegar is balsamic vinegar. It accounts for around 35 percent of all vinegar sold around the world. But unless you're very wealthy and very discerning, most of the balsamic you've consumed in your life probably wasn't true balsamic vinegar but instead the unaged Aceto Balsamico di Modena, a mixture of red wine vinegar, cooked grape must, and caramel. As far as shortcuts go, this is a perfect example of the trade-offs of doing things the easy way. While this cheaper version hits some of the notes of traditionally aged balsamic, much of its distinctive character is absent: viscosity, complex umami, and the essence of wooden barrels. Barrel-aging in wood infuses notes of caramel, vanilla, smoke, leather, and other distinct tones, depending on the variety of wood.

The traditional method for crafting balsamic vinegar takes place in both Modena and Reggio Emilia, neighboring cities in northern Italy. It requires five to nine wooden barrels made from different woods—mulberry, oak, juniper, cherry, ash, and acacia, to name a few—each built to different capacities. The barrels descend in size from 66 liters to 15 liters. At the outset, grape must that has been cooked down to caramelize and concentrate the sugars is fermented into a sweet wine

Aging vinegar in a barrel slowly reduces its volume over time through evaporation, and intensifies flavors while also imparting new ones.

by a plethora of different yeasts. From there, it's acidified into vinegar by AAB endemic to the environment. The largest barrels are filled with the vinegar, which is aged for a minimum of one year before being transferred to the next smallest barrel. Since wood is quasi-porous, water and some of the acetic acid are able to evaporate through the barrel, while larger aromatic compounds are not, leading to a milder, more concentrated flavor. The portion that disappears over time is known as the "angel's share" (a term also used in whiskey making), but the heavenly stuff is actually what gets left behind.

Only enough vinegar to fill the next smallest barrel is transferred. What was removed from the large barrel is replaced with freshly acidified grape must. The vinegar continues to proceed down the line to successively smaller barrels, and each barrel is topped up with liquid from the next larger barrel. For a traditional balsamic to be labeled DOP (Denomination of Origin of Production, a protected designation enforced by the European Union), the vinegar must be aged for at least twelve years. At the end of twelve years, a small portion of finished balsamic vinegar is drawn from the smallest barrel and finds its way to discriminating customers.

Balsamic vinegar production is laborious, to say the least. But in as little as three months, you'll see marked improvement after aging any vinegar in wooden barrels. Starting with a vinegar that already approximates the rich caramel flavors of balsamic is a good way of making up the time difference. We've had great success at Noma aging black garlic vinegar in wood. You can also get creative by infusing dried figs or plums into your favorite vinegar for a month before straining it and transferring it to a small barrel to age slowly.

Here in Denmark, we have few grapevines but plenty of elderberry trees, so when we set out to harness the qualities of a proper balsamic vinegar, we did so with what we had on hand. Read our recipe for Elderberry Wine Balsamic (page 201) for a look at an ongoing long-term project at Noma and an idea of how we try to learn from and adapt the traditions of faraway places.

169

Making perry vinegar begins with fermenting pureed
pears into alcoholic perry.

172

Perry Vinegar

Makes about 2 liters

4 kilograms sweet, ripe pears
1 packet (35 milliliters) liquid saison yeast
Unpasteurized pear vinegar, or another unpasteurized mild vinegar such as apple cider vinegar

In order to provide you with a good picture of both stages of vinegar fermentation, we'll begin with a recipe in which we first make alcohol through the fermentation of natural sugars, then ferment that alcohol into acetic acid with the help of AAB.

First, brewing the alcohol. Perry is pear cider—a sparkling, lightly alcoholic beverage that's as delicious chilled as it is warmed. There are dozens of varieties of pear; each will yield different perries and perry vinegars. In choosing the kind of pear you begin with, your guiding principle should be, *Would I want to drink the juice from this pear?* If the answer is yes, then by all means, rot on.

The skin of the pears hosts enough wild yeast to ferment them all on their own, but wild fermentation is always a gamble—you can never be sure what flavors will emerge, and the timeline is less predictable. This is more than fine in some cases, but since the perry will continue on to a secondary fermentation, we want a little more certainty about its flavor and alcohol content, so we'll depend on a yeast starter. The varieties of yeast available to ferment your perry are as varied as the pears themselves. (If you start multiplying the different variables in fermentation, you'll get a sense of just how expansive the flavor possibilities are.) Any well-stocked home-brew shop will be able to guide you to a yeast that will fare well with your pears. Stay away from baker's yeast, which will make your perry taste, well, bready. Here at Noma, we have

173

These are Conference pears, but more important than the varietal is that the fruit you choose is ripe and sweet.

a penchant for saison yeast, which is actually a blend of two different strains working side by side: *Brettanomyces* and *Saccharomyces*. We find it creates a great bouquet during fermentation, with no trace of bitterness.

Equipment Notes

For the first fermentation, you'll need a food-safe plastic bucket with a lid, airlock, and rubber stopper. You can find it at any home-brew shop. Look for a size of bucket that will hold the ingredients with about 15 percent of its volume to spare. You will also need a cider press or chinois to squeeze out the fermented liquid. You can perform the secondary fermentation in the same bucket, or use a smaller 3-liter wide-mouthed mason jar. Either way, you'll need cheesecloth or a clean kitchen towel, along with rubber bands to secure it to the top of the vessel.

Our quick method of producing vinegar requires an air pump and air stone, which you can find at a home-brew shop or pet store. Read the recipe for full details.

We recommend that all your equipment be thoroughly cleaned and sanitized (see page 36).

In-Depth Instructions

You need very sweet and ripe pears to make a decent perry. A crunchy d'Anjou pear might make for a nice snack, but it doesn't have a high enough sugar-to-fiber ratio to yield the alcohol content we're aiming for. Varieties like Bosc or Conference pears, which tend to sweeten significantly as they ripen, make for great perry.

The first stage of vinegar production is to use yeast to convert the sugar in fruit into alcohol. Find a bucket that will leave 15 percent of its volume after the pears are in. A 5-liter bucket is perfect here.

Stem the pears (you can leave the seeds in) and dice them into manageable pieces. Blend them into a rough puree in a food processor. It doesn't have to be completely smooth; just blend until you no longer see individual chunks of fruit.

Place the pear mash into the fermenting bucket. Add the yeast and mix, folding the fruit over to ensure that the yeast is well distributed. Snap the bucket lid shut, ensuring that it's airtight, then fill the airlock with water and insert it into the rubber stopper. (If you've never home-brewed before and this is hard to visualize, just ask the clerk at your home-brew shop or watch a video online. It's much easier than it sounds.)

Move the bucket to a spot that's slightly cooler than room temperature—about 18°C/64°F is ideal. Fermenting in warm temperatures can impart murky, musty tones to the perry. Ferment the perry for 7 to 10 days, depending on how much residual sweetness you want. Let taste be your guide. During fermentation, open the lid every day and stir the contents with gloved hands or a sterile spoon. There won't be any juice for you to taste in the early stages, but dipping a spoon into the pear mash will tell you everything you need to know. As the mixture ferments, the lid will puff up and the airlock will occasionally gurgle. This is caused by the carbon dioxide produced by the yeast, and it's perfectly normal. We don't advise taking the perry all the way to complete alcoholic fermentation (14 to 16 days), as you want some residual sugars to balance the flavor of acetic acid. If you find that your perry has fermented too far, you can simply add some fresh, strained pear juice to dilute it. It will be easier to adjust the balance of sugar at this point than later.

Once the pears have finished fermenting, you'll need to press the mash for its juice. At Noma, we do this with a cider press—basically a perforated metal or wooden drum that squishes juice from the fruit with a hand crank. You put the fermented mash in a cloth bag and place the bag in the drum. Turn the crank and out comes the juice via a spout at the base.

175

Perry Vinegar, day 1

Day 7

Day 14

If you aren't lucky enough to own a cider press, pushing the mash through a good old-fashioned chinois lined with cheesecloth will do the trick. Some fruit will pass through to the other side, so strain it again through a fine-mesh sieve or cheesecloth, but no need to get obsessive about straining. Viscosity isn't your enemy. A thicker juice makes wonderful perry with great mouthfeel and body.

So now you've got perry. And while this is a chapter about vinegar, you could chill this down and enjoy it right away, or warm it up and add some mulling spices, or transfer it to swing-top bottles and let it ferment further in the fridge into sparkling perry. However, the following step will make the last option impossible, so decide now if you want to turn your perry sour.

We don't want yeasts interfering with the flavor of the vinegar or continuing to ferment sugar into alcohol, so we kill them off. Transfer the strained perry to a pot with a lid and heat it to approximately 70°C/158°F—steaming but beneath a simmer. Cover the pot and hold it at that temperature for 15 minutes, stirring occasionally, then pull it off the heat and allow to cool to room temperature.

If you were to pour the perry into a couple of mason jars, cover them with cheesecloth, and leave them on the counter, you'd eventually have vinegar. We'll call that the long method—you'll be waiting somewhere around 3 to 4 months for the juice to acidify properly through wild fermentation.

To speed things up and give us more control, we do two things. First, we backslop (see page 33): Weigh the perry, then measure out 20% of that weight in unpasteurized pear vinegar (or a similar unpasteurized vinegar). For example, if you ended up with 1.8 kilograms perry, add 360 grams vinegar.

The second step is to aerate the vinegar. AAB need oxygen to function, and the long method does nothing to facilitate that. Begin by selecting the right fermentation vessel. You want something with lots of surface area, but nothing metal. You can use the same bucket you used for fermenting the pears, or switch to a 3-liter jar with a wide mouth. Pour the backslopped perry into the vessel. Wearing gloves, place the air stone into the liquid, making sure it rests on the bottom of the container. Snake the hose out the top of the vessel to the air pump and cover the vessel with cheesecloth or a breathable kitchen towel. Secure the cloth with a rubber band but be careful not to impede the flow of air through the hose. However, fruit flies absolutely adore the smell of vinegar—they're also known as "vinegar flies" in some regions—so it's important to make sure the seal on your cloth is unbroken. If there's a gap where the tube exits the bucket, use a piece of tape to shut it. Plug in your air pump and leave the perry to ferment at room temperature.

With constant aeration, you'll be able to turn the vinegar around in 10 to 14 days. Start tasting the vinegar daily after a few days. If the taste of alcohol is still noticeable, the vinegar needs to ferment further. You could use a pH meter or pH strips to test how acidic your vinegar is—a pH range of 3.5 to 4 is usually just right—but in all honesty, we find taste to be a better guide. Sugar, viscosity, and the flavor of your vinegar can all affect the perception of acidity on your tongue. A mechanical measurement may not necessarily lead to the product you want.

Once finished, strain your perry vinegar and store it in capped bottles in the fridge to keep the flavors as fresh as possible, though the vinegar is perfectly shelf stable as long as it's not exposed to air. If you notice any sediment at the bottom of the bottle, you can either shake the vinegar before using or, if you'd prefer a clear vinegar, gently pour it off into a fresh vessel, leaving the sediment behind (what we call "racking").

1. Dice the pears and blend them into a coarse puree.

2. Place the pear mash into a fermentation vessel, add the yeast, and cover with a lid and an airlock.

3. Allow the mixture to ferment for 7 to 10 days.

178

4. Press the mash to harvest the perry.

5. Transfer the perry to a new vessel, backslop with unpasteurized vinegar, and set up an air stone and pump.

6. Ferment until sufficiently soured, 10 to 14 days. Strain and bottle the finished vinegar, and store in the refrigerator.

Suggested Uses

Perry Vinaigrette

Perry vinegar has a light and delicate sweetness that yields the nicest vinaigrette of any vinegar in our repertoire. Whisk together 3 parts good olive oil, 1 part perry vinegar, and a small dollop of grainy mustard. Season with salt, and you've got all you need to elevate fresh salad greens, blanched wax beans, or lightly sautéed kale.

Pear Hollandaise or Béarnaise

Because perry vinegar doesn't have the same up-front harshness of your average white wine vinegar, it can stand alone as the base for sauces like hollandaise or béarnaise, where many classic recipes call for white wine vinegar diluted with white wine. Measure 250 milliliters perry vinegar into a small pot with a sliced shallot and a dozen peppercorns. Reduce the liquid by about two-thirds, then strain. Transfer the reduction to the top of a double boiler, add 3 egg yolks, then cook and whisk until the sauce thickens and falls off the whisk in ribbons. Season with salt and a bare pinch of cayenne pepper.

To double down on the fruitiness, brunoise firm but sweet pears and let them macerate in 250 milliliters perry vinegar for a couple of hours. Drain the pears, reserving the vinegar. Use the vinegar to make the reduction for the sauce, and fold the pear brunoise into the sauce. It's a vibrant and full-bodied sauce that you could just as easily serve alongside grilled hanger steak as with a bowl of barely cooked peas.

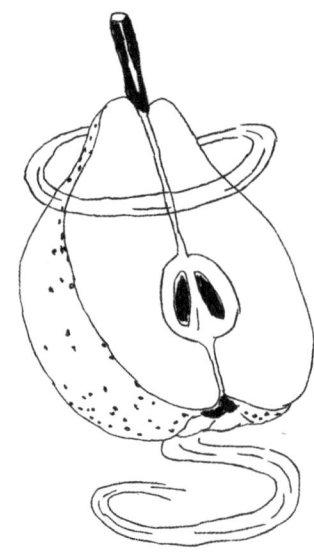

For More Accurate Sweetness, Use a Refractometer

If you want more precision in your measurements, use a refractometer. As explained in the kombucha chapter (page 118), a refractometer gives the sweetness of a solution in degrees Brix by measuring the refraction of light affected by the amount of sugar in water. As the perry ferments and the yeast turns sugar to alcohol, the Brix decreases.

To get an accurate reading with your refractometer, light needs to be able to pass through the liquid. In a situation like this, with chunks of fermented fruit, you'll need to squeeze some fruit to get a small portion of juice and then strain it through a fine-mesh sieve or cheesecloth. While the resulting liquid will still be a bit cloudy, it will give you a more accurate reading. Take an initial reading of the sugar content before you begin the ferment, then check it again every day or two. As the perry ferments, plug your measurements into the chart opposite, which converts the difference in Brix into alcohol by volume (ABV). The chart gives an approximate ABV for various changes in Brix for ferments held at a temperature of 20°C/68°F. You'll be ready to make vinegar once your perry is in the neighborhood of 6% to 7% ABV.

(Note that brewers often use a different measurement called "specific gravity," which is tied to the change in density of a liquid, to track the change in sugar content in a ferment. Because we end up fermenting products with fruit fiber in them, specific gravity is harder to measure accurately. It also requires a large sample size, usually a few hundred milliliters, which isn't feasible for some of our ferments, especially if you only want a quick check-in.)

Decrease in °Bx from starting point	Approximate % ABV
0	0
0.58	1
1.15	2
1.73	3
2.3	4
2.88	5
3.45	6
4.03	7
4.6	8
5.18	9
5.57	10
6.33	11
6.9	12
7.48	13
8.05	14
8.63	15
9.2	16
9.78	17
10.35	18
10.93	19
11.5	20

Plum wine fermenting into vinegar.

182

Plum Vinegar

Makes about 2 liters

4 kilograms ripe plums, rinsed, pitted, and cut into eighths

1 packet (35 milliliters) liquid saison yeast

Unpasteurized plum vinegar or another unpasteurized mild vinegar, such as apple cider vinegar

This is another two-stage vinegar where we convert the sugar in fruit into alcohol, then the alcohol into acetic acid. Black, purple, or deep red plums will yield the prettiest vinegars, and a mix of varieties will create more complex flavors.

The in-depth instructions for Perry Vinegar (page 173) serve as a template for this vinegar. We recommend you read that recipe before starting in on this one.

Place the cut plums into the bucket. Add the yeast and mix, folding the fruit over to ensure that the yeast is well distributed. Plums have a higher water content than pome fruits like pears or apples and as such don't need to be pureed into a mash to ferment well—they'll liquefy on their own. Snap the bucket lid shut, ensuring that it's airtight, then fill the airlock with water and insert it into the rubber stopper.

Ferment the plums in a cool room, stirring the mixture once every day, for a total of 8 to 10 days, until they're noticeably alcoholic without having lost all their sweetness.

Press the plums using either a cider press or your hands and a cheesecloth-lined chinois; then strain once more. You now have plum wine. Measure and note the weight of the wine, then transfer it to a pot with a lid. Heat the wine to 70°C/158°F—steaming, but below a simmer—then cover the

183

Plum Vinegar, day 1

Day 7

Day 14

pot and hold it at that temperature for 15 minutes. Pull it off the heat and allow it to cool to room temperature.

Pour the wine into the secondary fermentation vessel—either the bucket you used to ferment the plums or an 8-liter wide-mouthed jar. Backslop with 20% of the wine's weight in unpasteurized vinegar. Place the air stone in the liquid so it rests on the bottom of the vessel and run the hose out the top to the air pump. Cover the vessel with cheesecloth or a breathable kitchen towel and secure with a rubber band. Tape over the gap left by the hose and turn on the pump.

Ferment the plum vinegar for 10 to 14 days, tasting frequently during the home stretch. When all the alcoholic flavor is gone and the vinegar is nicely acidic but still fruity, strain it through cheesecloth. Store in capped bottles in the refrigerator to keep the flavors as fresh as possible, though the vinegar is perfectly shelf stable as long as it's not exposed to air.

Suggested Uses

Marinade for Roasted or Grilled Meat

Plum vinegar works well as a base for meat marinades. Sear a few oxtails in a pan over high heat to caramelize their exterior. Combine equal parts vinegar, beef stock, Beef Garum (page 373), and olive oil. Set aside a few spoonfuls of the marinade and use the rest to soak your oxtails in a plastic bag in the fridge for 2 hours. Transfer the oxtails to a roasting pan with the usual suspects—aromatic vegetables and your favorite herbs—cover tightly with foil, and cook slowly in a low oven (160°C/320°F) for a few hours. Once the meat is tender, pull it off the bone, dress with the reserved marinade, and season to taste.

Or use the same combination to marinate beef or pork spareribs. Grill them slowly over medium heat until crusty on the outside and tender within. The marinade infuses the meat with tartness and umami, like good barbecue should have, without the stickiness or sweetness of traditional sauces.

Christmas Cabbage

In Scandinavia, during the cold days around Christmas, cabbage is on everyone's mind. Core and slice a head of red cabbage as thinly as you can, then sauté it in a healthy amount of duck fat (about 100 grams for a head of cabbage). Add 200 milliliters plum vinegar per head of cabbage, top with a lid, and simmer slowly for 2 hours, stirring and scraping the pan occasionally. You don't want the liquid to dry up, but rather reduce and merge with the softened cabbage to form a rich compote. Once you're there, season with salt and serve. Or go the extra mile and make Roasted Chicken Wing Garum (page 389), and add a few spoonfuls to up the umami.

Red cabbage is a ubiquitous presence on Scandinavian tables in the winter. We like to cook it down very slowly with plum vinegar and roasted chicken wing garum.

Above: Celery juice ferments into a surprisingly fresh, vegetal vinegar.
Opposite: Check the pH of vinegar against the legend that comes with the test kit.

Celery Vinegar

Makes about 2.5 liters

3 kilograms celery, broken into stalks
    and rinsed to remove any grit
Unpasteurized celery vinegar or
    another unpasteurized mild vinegar,
    such as apple cider vinegar
96% ethanol (neutral grain spirit)

The first time you taste celery vinegar, you'll wonder how you've lived without it. The flavor is surprisingly versatile. The delicate green notes of celery play well with vegetable salads, or as a kicker in gazpacho, or mixed with walnut oil to dress sliced apples.

This fermentation method differs from the two-stage fermentations we employ for Perry Vinegar (page 173) and Plum Vinegar (page 183). Rather than fermenting our own alcohol, we add ethanol to vegetable juice to provide fuel for the acetic acid bacteria (AAB). Still, the in-depth instructions for Perry Vinegar have useful information about our quick method and may be worth reading before starting in on this one.

Equipment Notes

You'll need a juicer, a food-grade plastic bucket or wide-mouthed jar with at least a 3-liter capacity, and an air pump and air stone. You'll also want cheesecloth or a breathable kitchen towel, along with large rubber bands to secure it to the top of your fermentation vessel. We recommend that you wear sterile gloves when working with your hands, and that all your equipment be thoroughly cleaned and sanitized (see page 36).

In-Depth Instructions

Juice the celery. (The fibers in celery have a tendency to gum up the blade, so you may need to clean out the juicer once or

Celery Vinegar, day 1

Day 7

Day 14

twice during the process.) Strain the juice through a fine-mesh sieve. Weigh it and pour it into the fermentation vessel.

Backslop the juice with 20% of its weight in unpasteurized vinegar. For example, if you end up with 2 kilograms juice, you'll backslop with 400 grams unpasteurized vinegar.

Next, you'll need to add alcohol to fuel the AAB. To determine the amount of alcohol needed, add together the weight of the juice and the vinegar and calculate 8% of that weight. For example, if the total weight of juice and vinegar is 2.4 kilograms, you'll add 192 grams of 96% ethanol. Add the ethanol to the fermentation vessel.

Your base mixture is now ready for fermentation, using the same setup as described for perry vinegar. One-stage vinegars like celery vinegar need to be pushed through their fermentation as quickly as possible to maintain the quality of the juice and integrity of the flavors. Passive, long fermentation is not an option here.

Place the air stone in the liquid so that it rests on the bottom of the vessel and run the hose out the top to the air pump. Cover the vessel with cheesecloth or a breathable kitchen towel and secure with a rubber band. Tape over the gap left by the hose and turn on the pump. If you find your vinegar pushing foam out the top of its container, the pump is likely working too hard. You can add a small plastic valve to the line of the hose to regulate its flow, or simply ferment in a bigger vessel. Otherwise, stop the pump and stir the foam back into the vinegar whenever it builds up too much. Either way, an excess of foam usually subsides after the first few days.

You'll notice that the alcohol in a one-stage vinegar tastes more pronounced than in a two-stage fermentation. Pure ethanol has a conspicuous flavor that lingers until it's been completely fermented away. Near the end of the fermentation, it's possible that you'll pick up faint notes of varnish or nail polish remover. With a bit more time and aeration, those scents will dissipate.

188

If you can't find 96% ethanol, use a neutral spirit such as vodka. However, since the ABV of vodka is much lower (80 proof, or 40%, rather than 96%), you'll have to adjust the recipe to account for the extra water. First, increase the amount of vinegar we use to backslop, adding 23.4% by weight rather than 20%. In the example used in the celery vinegar recipe, if we have 2 kilograms celery juice, we'd add 468 grams vinegar.

Next, we'll also have to add more alcohol to make up for the difference in ABV: 20% by weight rather than 8%. For 2.468 kilograms of the celery juice and vinegar mixture, we'd add 494 grams vodka.

Keep in mind that adding all this additional liquid dilutes the juice and its flavor. The final vinegar will also be more dilute, but definitely recognizable as celery vinegar.

(Like all good experimentalists, you may want to try a more flavorful alcohol. Feel free, but try it this way first.)

You'll also learn that the vibrant chlorophyll of green vegetables fades when making vinegar, as acetic acid degrades the pigment molecules and changes the color. Don't fret about the dull olive hue of your vinegar—the finished flavor will more than make up for it.

Ferment the vinegar for 10 to 14 days, or until all the flavor of the alcohol has vanished and the vinegar has acidified while retaining the flavor of fresh celery. Strain it through cheese-cloth. Store in capped bottles in the refrigerator to keep the flavors as fresh as possible, though the vinegar is perfectly shelf stable as long as it's not exposed to air. If you notice sediment at the bottom of the bottle, just shake before using; otherwise, you can gently pour the vinegar off into a fresh vessel, leaving the sediment behind (what we call "racking").

Suggested Uses

Cucumber Soup

Cut a couple of English cucumbers into manageable pieces and blend them until smooth with a pinch of salt and a couple of teaspoons of Grasshopper Garum (page 393) to season, before passing through a fine-mesh sieve. Stir in about 150 milliliters celery vinegar to lift everything with green, piquant acidity. Chill the soup in an ice bath and enjoy as is, or dice up your favorite summer vegetables and add them for a refreshing starter.

Celery-Herb Vinegar with Fresh Cheese

Celery vinegar, with its bright vegetal flavor, is a perfect starting point to create infused vinegars. Take 500 milliliters celery vinegar and blend it with 100 grams fennel tops, parsley leaves, or whatever sweet herb you like. Let the mixture infuse for about 5 minutes, then strain. You'll have a vibrant green, herbaceous vinegar with the pure flavor of fresh herbs. From there, you can dress a bowl of fresh ricotta or mozzarella with olive oil, sea salt, and red pepper flakes, then finish with a drizzle of celery-herb vinegar just before serving.

The lightly sweet juice of butternut squash doesn't
have enough sugar to ferment into the required
amount of ethanol. Adding neutral grain spirit allows
it to be fermented into vinegar.

190

Butternut Squash Vinegar

Makes about 2 liters

4 kilograms butternut squash
Unpasteurized butternut squash
    vinegar, or another unpasteurized
    mild vinegar such as apple
    cider vinegar
96% ethanol (neutral grain spirit)

This vinegar is by far the most adaptable of all the fermentation recipes we employ at Noma. It has a nice acidic kick but doesn't sport an overt sharpness—the butternut squash's almost creamy sweetness makes you think the acidity level is lower than it is. It can practically be used as a sauce as is.

The in-depth instructions for Celery Vinegar (page 187) serve as a template for this recipe, and are worth reading before starting in on this one. You may also want to read the in-depth instructions for Perry Vinegar (page 173), which contain useful information for all the vinegar recipes in this chapter.

Wash the squash, cut them in half, seed them, and cut them into manageable pieces, leaving the skin on. Wearing gloves, put the squash through the juicer. Strain the juice through a fine-mesh sieve. Weigh it and pour it into the fermentation vessel.

Backslop the juice with 20% of its weight in unpasteurized vinegar. Calculate 8% of the total weight of the juice and vinegar and add that much ethanol. (If you can't acquire pure ethanol, adjust your recipe to use an 80-proof spirit like vodka according to the instructions on page 189.)

Place the air stone in the liquid so it rests on the bottom of the vessel and run the hose out the top to the air pump. Cover the vessel with cheesecloth or a breathable kitchen towel and

Butternut Squash Vinegar, day 1

Day 7

Day 14

secure with a rubber band. Tape over the gap left by the hose and turn on the pump.

Ferment the butternut squash vinegar for 10 to 14 days, tasting frequently during the home stretch. If the juice foams during the first few days, turn off the air pump for a bit or stir the foam back into the juice. When you can no longer taste the alcohol and the vinegar is enjoyably acidic, strain it through cheesecloth. Store in capped bottles in the refrigerator to keep the flavors as fresh as possible, though the vinegar is perfectly shelf stable as long as it's not exposed to air. The bright orange color will fade over time.

Other quick vinegars (juice + ethanol + backslopped vinegar):

- Beet
- Bell pepper
- Black currant
- Carrot
- Cauliflower
- Celeriac
- Cucumber
- Fennel
- Jicama
- Kelp and katsuobushi dashi
- Quince
- Sweet potato
- White asparagus

Suggested Uses

Slow-Cooked Carrots

Find some nice carrots—not the monstrous horse carrots you might use for stock. Peel and slice them however you like—into thin strips or on a bias—or leave them whole. Melt a big knob of butter in a pan over low heat and slowly caramelize the carrots in a single layer—as in, really slowly, with the butter gently foaming and bubbling. Turn the carrots every 6 or 7 minutes for anywhere between 30 and 50 minutes (depending on what your idea of low heat is). If you've done it correctly, the carrots should take on a caramelized color and texture reminiscent of golden raisins. When they're almost there, turn the heat up a tiny bit and add a touch of salt and one spoonful

Deglaze slow-roasted carrots with butternut squash vinegar.

Button chanterelle mushrooms soaked in butternut squash vinegar for a day (or for up to a year) make outstanding pickles.

of squash vinegar for every two or three carrots. You want just enough liquid to coat the carrots lightly, giving them a kick of acidity and another dimension of flavor. This technique works great for other vegetables, too—parsnips, turnips, rutabaga, pumpkin, or anything that takes well to slow cooking.

Quick Pickles

You can try this with any crunchy fruit or vegetable you'd enjoy eating raw, but let's use a cucumber as an example. Slice the cucumber into thin (3-millimeter / ⅛-inch) coins and season the slices lightly with salt, letting them marinate in a bowl for about 10 minutes before covering with butternut squash vinegar. Stir everything around to ensure even coverage, adding a bit of red pepper flakes for heat, if you like. Do this about an hour before dinner and they'll be perfectly pickled by the time you're sitting down to eat.

One other thing we love to pickle with butternut squash vinegar is chanterelle mushrooms. In a skillet, lightly sauté cleaned mushrooms in as little oil as possible, making sure they're cooked but not mushy. Let them cool down on a plate, then transfer to a glass jar. Cover with twice their volume in vinegar (they'll soak up a fair bit of it) and seal the jar tightly. They'll be delicious by the next day, but the pickles will last several months in the refrigerator. If you take the canning process a step further and process the jars using the boiling water bath method, the pickles will last even longer—6 months to a year in a cool dark place. They're a perfect condiment alongside roast chicken or fish and make a great gift.

Sautéed Shrimp

The next time you're sautéing peeled shrimp, add a splash of equal parts butternut squash vinegar and shrimp garum, just as the shrimp begin to turn opaque. (If you haven't gotten around to making the Rose and Shrimp Garum on page 381, substitute a 1:1 ratio of Worcestershire and fish sauce.) The liquid will deglaze the pan and coat the shrimp as they caramelize—delicious stuff.

193

Some of the alcohol needs to be burned off before
whiskey can be fermented into vinegar.

194

Whiskey Vinegar

Makes about 2 liters

1.5 kilograms + 350 grams 80-proof
  whiskey
400 grams unpasteurized apple
  cider vinegar
Water

In this third method of vinegar making, we start with an alcoholic base liquid and reduce the ABV from 40% to around 8%. We accomplish this by burning off nearly all the alcohol in the spirit, then adding a fresh dose to bring the ABV back up. The trick is to find an alcohol that has enough character to remain delicious even after being cooked and diluted.

We tested whiskey vinegar in advance of our pop-up restaurant in Sydney as an acknowledgment of Australia's whiskey-making tradition. The vinegar didn't end up making it into any dishes, which is just the nature of menu development at Noma, but we loved it, and thought it worthy enough to include here. It's got a bit more punch than some of the other vinegars in this book, and it works well with flavorful meats, especially if you offset the acidity with a little sweetness.

The in-depth instructions for Perry Vinegar (page 173) contain useful information for all the vinegar recipes in this chapter. We recommend you read that recipe before starting in on this one.

Equipment Notes

You'll need a food-grade plastic bucket or wide-mouthed jar with at least a 3-liter capacity, and an air pump and air stone. You'll also want cheesecloth or a breathable kitchen towel, along with large rubber bands to secure it to the top of your

195

Whiskey Vinegar, day 1

Day 7

Day 14

fermentation vessel. We recommend that you wear sterile gloves when working with your hands, and that all your equipment be thoroughly cleaned and sanitized (see page 36).

In-Depth Instructions

Preheat a large two-handled pot with a lid over medium heat. Be sure to use as deep a pot as possible, as the whiskey can boil over, something you must avoid at all costs. The pot shouldn't be smoking, but it should be very hot, the idea being that you'll flash-boil the alcohol out of the whiskey. Make sure there is nothing flammable above your stovetop and that there is no heat-sensitive fire alarm nearby.

Once the pot is up to temperature, carefully and quickly pour in about 500 grams of the whiskey. It will boil instantaneously, and could ignite right away. If it doesn't, use a grill lighter or a long match to ignite the alcohol. *Please exercise caution*, as lit alcohol can cause severe burns, and the flames are difficult to see but can grow quite high. If at any time you feel nervous about the size of the flame, turn off the heat and snuff out the flame by covering the pot with a tight-fitting lid.

Once the flames have subsided, add the next 500 grams of the whiskey. Repeat until you've burned off all the alcohol from the 1.5 kilograms whiskey. Remove from the heat. The volume will be greatly reduced, as you've also burned off more than 40 percent of its volume. (Note that it's very difficult to remove 100 percent of the alcohol, but that's fine for our purposes.) All the heavy flavor particles that the whiskey inherited from its time barrel-aging are now concentrated in the reduced liquid.

Add as much water as necessary to bring the liquid up to 1.25 kilograms total. To this base you'll add the remaining 350 grams untouched whiskey and the 400 grams unpasteurized apple cider vinegar. This yields a liquid with an approximate total alcohol content of 8% and enough acetic acid bacteria (AAB) to kick-start the acidification process.

Combine the reduced whiskey, untouched whiskey, and apple cider vinegar to begin fermentation.

Beef garum mixed with whiskey vinegar forms our version of the Vietnamese sauce nuoc cham.

Transfer the liquid to the fermentation vessel. With whiskey you're not at risk of the flavors changing over a long passive fermentation, which would occur with fruit or vegetable vinegars, so you could simply cover it with cheesecloth, secure the cheesecloth with a rubber band, and leave it at room temperature for 3 to 4 months to become vinegar.

For more reliable and quicker results, place the air stone in the liquid so it rests on the bottom of the vessel and run the hose out the top to the air pump. Cover the vessel with cheesecloth or a breathable kitchen towel and secure with a rubber band. Tape over the gap left by the hose and turn on the pump.

Ferment the whiskey mixture for a little less time than other vinegars, maybe 8 to 12 days, tasting frequently during the home stretch. The lack of any sugar in the whiskey can lead to a slightly dry vinegar that can taste flat if you don't pull it a little early. In a well-balanced whiskey vinegar, the residual alcohol content won't be too pronounced, and the whole thing will sit on your palate with roundness and warmth.

There's no need to strain this vinegar, unless you notice some sediment. Store in capped bottles in the refrigerator to keep the flavors as fresh as possible, though the vinegar is perfectly shelf stable as long as it's not exposed to air.

Suggested Use

Whiskey Vinegar Sauce

The classic Vietnamese dipping sauce nuoc cham—made from fish sauce, lime juice, and sugar—is a perfect amalgamation of acidity, sweetness, and funk that we can try to emulate using our own ferments. Mix 4 parts whiskey vinegar with 1 part honey. Season with a splash of Beef Garum (page 373). You'll end up with the ideal dipping sauce for red meats like duck, quail, or aged beef, as well as cooked and raw greens or root vegetables.

Gammel Dansk Vinegar

Makes about 2 liters

400 grams Gammel Dansk
1.185 kilograms water
350 grams unpasteurized apple
    cider vinegar

Gammel Dansk is the classic herbal liqueur of Denmark. At Noma, we've used it in desserts and always have a bottle or two floating around in the lounge for those who like a bracing shot of bitters after dinner. The alcohol content of Gammel Dansk is lower than that of a spirit like whiskey. As such, we approach lowering its ABV to a percentage suitable for vinegar production in a straightforward manner: We dilute it. The flavor's so strong to begin with that watering it down doesn't diminish its effect.

The in-depth instructions for Perry Vinegar (page 173) contain useful information for all the vinegar recipes in this chapter. We recommend you read that recipe before starting in on this one.

Add the Gammel Dansk, water, and vinegar to the fermentation vessel and stir to combine them well. Place the air stone in the liquid so it rests on the bottom of the vessel and run the hose out the top to the air pump. Cover the vessel with cheesecloth or a breathable kitchen towel and secure with a rubber band. Tape over the gap left by the hose and turn on the pump.

Ferment the Gammel Dansk vinegar for 8 to 12 days, or until sufficiently acidified. There's no need to strain this vinegar, unless you notice some sediment. Store in capped bottles in the refrigerator to keep the flavors as fresh as possible, though the vinegar is perfectly shelf stable as long as it's not exposed to air.

Gammel Dansk Vinegar, day 1

Day 7

Day 12

Suggested Use

Bitters as a Booster

Gammel Dansk vinegar is a rather intense character, but it's fun to play around with its bitterness and acidity, as these two attributes work well together and tend to soften each other's effect. It's not a vinegar that you want to set out as a condiment on the kids' table, but you can certainly sneak a spoonful in here and there for an unexpected boost of complexity and edge. The next time you make boeuf bourguignon, for example, a few splashes of Gammel Dansk vinegar—just enough so that you barely register it on your palate—will subtly elevate your stew's savory richness. In smaller doses, this vinegar can improve a Waldorf salad, or creamier dressings made with mayonnaise or crème fraîche.

The addition of fresh elderberries to elderflower wine gives a boost in flavor for the years-long aging process. Elderberries are mildly poisonous and can upset stomachs if eaten raw. Through cooking or fermentation, they're rendered safe for consumption.

Elderberry Wine Balsamic

Makes about 5 liters (before aging)

1.15 kilograms sugar
1.15 + 1.7 kilograms water
500 grams elderflower blossoms,
    stems removed
1 packet (35 milliliters) liquid
    saison yeast
1 kilogram unpasteurized apple
    cider vinegar
600 grams ripe elderberries,
    stems removed

There are few flavors more representative of Scandinavia than elderflower, and few ingredients that play a bigger part in the experience of working at Noma. Picture twenty-five cooks situated around long tables sorting through garbage bag upon garbage bag stuffed with shrubby branches, fighting off allergies while picking tiny flower after tiny flower at an ever-accelerating pace to process 60 kilos before it's time to clean up—to start it all again the next day.

Elderberry vinegar has been a staple in our kitchen for some time. This elderberry wine balsamic, however, is an ongoing experiment. We'll have no idea of the final flavor for another ten years. If you want to join us in our trial run, this recipe shows how we did it.

Equipment Notes

You'll need a 5-liter food-safe plastic bucket fitted with an airtight lid, a rubber stopper, and an airlock for the first-stage fermentation of the elderflower wine. You can reuse the same container for the secondary fermentation, or use a 5-liter wide-mouthed jar. You'll also need to procure a 5-liter wooden barrel for aging, and more barrels of decreasing size should you decide to age your vinegar for very long periods of time. A refractometer is optional. We recommend that you wear sterile gloves when working with your hands, and that all your equipment be thoroughly cleaned and sanitized (see page 36).

201

Elderberry Wine Balsamic, day 1

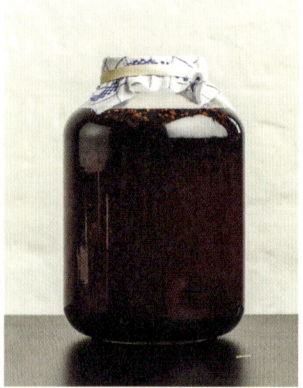

Day 7

Day 14

Balsamic vinegar isn't normally a part of the Noma pantry, but it's a vital pillar of gastronomy and a beloved ingredient in kitchens around the world. The process of making balsamic vinegar touches on some of the most fascinating aspects of fermentation and aging, so we've developed the following recipe for our own version.

In-Depth Instructions

Begin the years-long process of producing balsamic vinegar by making elderflower syrup. Bring the sugar and 1.15 kilograms of the water to a boil in a large pot, stirring to dissolve the sugar, then remove from the heat. While the syrup is coming to a boil, put the elderflowers in a clean heatproof container. Pour the syrup over the elderflowers and allow to cool to room temperature. Place a couple of sheets of plastic wrap in direct contact with the surface of the liquid—the elderflowers have a tendency to float and the plastic will help keep them submerged—then transfer the syrup to the fridge, covered, to infuse for 2 weeks.

Strain the syrup through a fine-mesh sieve into the 5-liter bucket, pressing on the elderflowers for maximum extraction. Add the remaining 1.7 kilograms water, which will bring the sugar content down to 30°Bx (from a starting point of 50°Bx). If you have a refractometer, take an exact measurement of the Brix level, which will serve as a point of reference for how far you take the initial fermentation. Add the yeast to the diluted elderflower syrup and stir with a clean spoon. Snap the bucket lid shut, ensuring that it's airtight, then fill the airlock with water and insert it into the rubber stopper.

Move the bucket to a spot that's slightly cooler than room temperature—about 18°C/64°F is ideal—and allow it to ferment for 2 to 3 weeks. We want a good amount of residual sweetness left in the elderberry wine, and an alcohol level of 8% to 10% ABV. If you're using a refractometer and you've taken the initial reading, test the Brix again after the fourteenth day of fermentation. Use the chart on page 181 to convert the difference in your measurements to ABV.

Add liquid saison yeast to begin fermenting the sugars in the elderflower syrup.

Aging elderberry vinegar in a wooden barrel
will produce layers of nuance and complexity.
The longer it ages, the better.

Choosing an Aging Barrel

Sourcing a barrel can be quite a bit of fun. To age balsamic vinegar, you're looking for a wooden barrel, as the porousness of wood will allow the vinegar to evaporate over time. Traditionally, the barrels' bung holes aren't plugged closed, but rather covered with cloth to expedite evaporation without letting anything in. The interior of a wooden barrel is usually charred with an open flame, which creates many of the barrel's flavors in the form of volatile compounds like vanillin, along with tannins and terpenes.

The variety of wood you choose is up to you—they all have their own qualities. Starting off with a 5-liter barrel, it's not difficult to find decreasing sizes all the way down to 1 liter.

Our elderberry wine balsamic is currently aging in used Bruichladdich Scotch whisky barrels made from oak. Used barrels will impart an idiosyncratic flavor to your vinegar and can include anything from wine barrels to bourbon barrels to sherry casks. Regardless of whether you've acquired a used or brand-new barrel, you should fill it with water and let it soak for one day to swell the wood before adding your vinegar. This will ensure that the barrel is watertight.

Once you've reached the desired alcohol content, backslop with the apple cider vinegar and add the whole elderberries. Replace the lid and airlock with a piece of cheesecloth securely fastened with a rubber band. Leave the wine to ferment at room temperature for 3 to 4 months, stirring every few days with a clean spoon, as the berries tend to float. This is a vinegar that acidifies very well at a slow pace. At the restaurant, we've traditionally made it in this fashion, but if you'd like to get through the acidifying stage faster, feel free to employ an air pump and air stone as described throughout this chapter.

Once the vinegar has fermented to your liking, strain it through a fine-mesh sieve, pressing the berries against the sides for maximum extraction, and then again through cheesecloth. Using a funnel, transfer the vinegar to the barrel and cap the hole. Leave the barrel to rest in a cool room or basement, ideally around 18°C/64°F. The humidity of the environment will affect the rate of evaporation within the barrel. The drier the room, the faster it will dissipate. As we're seeking to age this vinegar for years, you'll want to mediate evaporation, so avoid overly dry environments.

Traditional balsamic is aged for a minimum of 12 years, but you'll notice a remarkable change in flavor after just one year. If you're in it for the long haul, you'll want to decant the barrel's contents into a smaller barrel after 12 months. As time goes by, the volume of vinegar will decrease through evaporation. By stepping down a barrel size each year, you'll maximize the amount of contact the vinegar has with the wood, meaning more flavor transfer. Select barrels that just fit the amount of vinegar you have remaining. This is our plan for the elderberry vinegar aging at Noma, as we try to develop its flavor and complexity over the long run.

205

Black Garlic
Balsamic

Makes about 5 liters (before aging)

500 grams Black Garlic (page 417)
3.375 kilograms water
1.125 kilograms sugar
1 packet (40 milliliters)
    Chardonnay yeast
Unpasteurized apple cider vinegar

At Noma, we're constantly seeking ways to use fermentation to reduce food waste of all kinds. The original version of this vinegar was a result of an abundance of black garlic skins left over from making a sort of fruit leather out of black garlic cloves. We've adapted this recipe to use the whole heads of garlic, skins and all. The in-depth instructions for Elderberry Wine Balsamic (page 201) serve as a template for this vinegar as well. We recommend you read that recipe before starting in on this one.

Cut the heads of black garlic in half laterally, skins and all, then slice each half vertically into 4 pieces. Place the water and sugar in a large pot and bring the mixture to a simmer, whisking to dissolve the sugar. Add the garlic, then lower the heat so that the liquid is barely steaming. Let it sit, covered, on the stove for 1 hour, then remove from the heat and allow to cool to room temperature. Once cooled, transfer the covered black garlic stock to the fridge to continue to infuse overnight.

Strain the stock through a chinois into a bowl, using a ladle to force out as much liquid as possible without pushing any of the garlic's flesh through. Pass the strained stock through a cheesecloth-lined sieve. Add the yeast to the black garlic stock, stir it with a clean spoon, and transfer to the fermentation vessel. Snap the bucket lid shut, ensuring that it's airtight, then fill the airlock with water and insert it into the rubber stopper.

Black Garlic Balsamic, day 1

Day 7

Day 14

Allow the stock to ferment in a cool room for 2 to 3 weeks. You now have black garlic wine. It should be noticeably alcoholic, while still having a good amount of residual sweetness.

Strain the black garlic wine through a cheesecloth-lined sieve and transfer it to the secondary fermentation vessel— either the same bucket you used to ferment the wine or a 5-liter wide-mouthed jar. Weigh the wine and backslop with 20% of the wine's weight in unpasteurized apple cider vinegar. Place the air stone in the liquid so it rests on the bottom of the vessel and run the hose out the top to the air pump. Cover the vessel with cheesecloth or a breathable kitchen towel and secure with a rubber band. Tape over the gap left by the hose if necessary and turn on the pump.

Allow the wine to ferment into vinegar at room temperature for about 14 days. Taste the vinegar to ensure enough acetic acid has been produced. Strain the vinegar and then transfer it to the wooden barrel to age and reduce. Leave the barrel to rest in a cool room or basement, ideally around 18°C/ 64°F. You can age this for years and it will continue to grow in character. After just one year, you'll notice a tremendous difference, but should you want to leave it for longer, decanting it into decreasing barrel sizes as you go, your patience will be rewarded.

Suggested Uses

Substitute for Chinese Black Vinegar

Black garlic vinegar, whether young or barrel-aged, can be used anywhere you'd use Chinese black vinegar. A ramekin of black garlic vinegar, embellished with nothing more than a drop of sesame oil and a splash of chili oil, makes for a near-perfect dipping sauce for dumplings or steamed buns. Crunchy greens like bok choy or gai lan, freshly steamed and still hot, beg to be tossed in a splash of black garlic vinegar and seasoned with furikake, the Japanese dried topping made of toasted seaweed, katsuobushi, and sesame seeds.

Black Garlic Vinegar and Ryeso Sauce

Another nominally Asian use for black garlic vinegar is a reinterpretation of Chinese black bean and garlic sauce. In place of the dried and fermented Chinese black beans (*douchi*), we'll use another ferment from the book: Ryeso (page 307). In a mortar and pestle, grind 100 grams ryeso until almost smooth, then add 50 grams black garlic vinegar, continuing to muddle the mixture until well incorporated. (Ideally, you'd use barrel-aged vinegar for its thicker viscosity, but if you're impatient and don't have a year to kill, you can just as easily reduce fresh vinegar in a pan by two-thirds before mixing it with the ryeso.) Season with salt and finish with grated fresh horseradish, which will step in for the warm heat of chili oil. This is a much different beast than the sauce you're used to, but it mimics the sweet malted notes of fermented black beans thanks to the slow caramelization that occurs in making both the ryeso and the black garlic. It serves as an amazing accompaniment to red meats, whether as a dip or slathered directly over top. If you find you're missing the telltale funk of the fermented black beans, don't hesitate to add a few drops of Squid Garum (page 385).

Muddle ryeso with black garlic balsamic
for a potent, umami-rich dipping
sauce or seasoning paste.

5.

Koji

—

Pearl Barley Koji 231

Citric Barley Koji 243

Sweet Citric Koji Water 246

Sparkling Citric Koji Amazake 249

Dried Koji and Koji Flour 253

Lacto Koji Water 259

Roasted Koji "Mole" 263

Koji Cure (Shio Koji) 265

The Magic Mold

"Any sufficiently advanced technology is indistinguishable from magic," said Sir Arthur C. Clarke. And some of the most remarkable technologies on earth are biological—systems blindly refined by chance and circumstance, built over eons. The natural world is an endless source of wonder, its infinite variation a bottomless well of discovery.

We find koji indistinguishable from magic—the best kind of magic, in fact, because anybody can wield it. To experience koji's brilliance for yourself, you simply need to pop some in your mouth and taste it.

Koji is a term that comes from Japan, where it refers to rice or barley that has been inoculated with *Aspergillus oryzae*, a species of fungus—a sporulating mold, to be exact—that grows on cooked grains in warm and humid environments. (In the English-speaking world, we apply the term *koji* interchangeably to the inoculated grains, the fungus, and the spores.)

Under the right conditions, when the microscopic spores of *Aspergillus oryzae* land on a suitable substrate like cooked barley or rice, the spores will sprout hyphae—branching fungal cells resembling wispy white roots. As the fungal cells multiply, the hyphae dig into the grains, spreading their tendrils as they grow to form a network known as a mycelium. What begins as a few patches of white fuzz grows over the course of two days to form a dense white mat that completely binds and encases the grains. The resulting "cake" of moldy grains is koji. After the first twenty-four hours it begins to release an intoxicating bouquet of smells, redolent of passion fruit and apricots. After forty-eight, the koji is sweet, fruity, and full of umami.

The chemicals responsible for koji's flavor and aroma are enzymes released by the hyphae as they grow into the grains, digesting the substrate externally, and absorbing the nutrients in order to fuel its metabolism. The fungus produces a flight of enzymes that break down starches (amylase), proteins (protease), and fats (lipase) into their constituent building blocks of simple sugars, amino acids,

Aspergillus hyphae spread into a visible web as they grow.

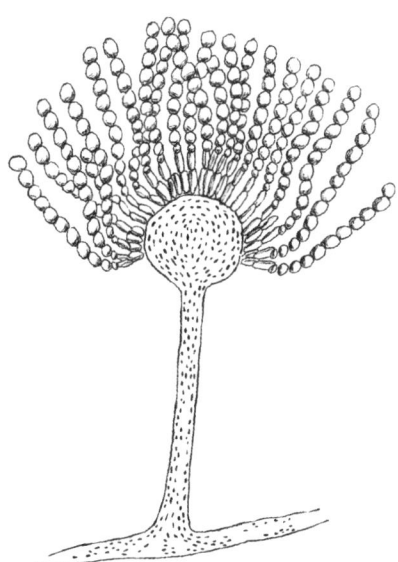

Aspergillus oryzae, the magic mold.

and fatty acids, respectively. (You can identify an enzyme by the suffix *-ase*; the prefix describes the substance it works on.)

Once you begin poking around the world of fermentation, you'll stumble into koji very quickly. It's nearly impossible not to encounter it, like going to Paris and not seeing the Eiffel Tower. Yet koji has only really made headway into Western kitchens in the past five or ten years. As for us at Noma, a trip to Japan in 2010 really opened our eyes to the possibilities that koji presents.

Ajinomoto—a company responsible for manufacturing all kinds of seasoning products, including much of the world's MSG—operates a research facility in Tokyo called the Umami Information Center. They do serious work there, researching umami and its uses; and while we had already been dabbling with koji in Copenhagen, after our trip to the Center we returned home fixated on furthering our umami work. Our test kitchen and fermentation lab spent weeks not really creating dishes, but rather exploring ways of extracting umami from things native to our part of the world. It quickly became obvious that koji was going to be the key to unlocking the "fifth taste" for us.

213

214

Unripe Macadamia Nuts and Spanner
Crab, Noma Australia, 2016

Slivers of unripe macadamia nuts rest in
a chilled, clear broth of Australian spanner
crab seasoned heavily with lacto koji
water and rose oil.

There are plenty of ingredients that are excellent at being
products unto themselves; many others function better
as cooking tools. Only a handful are good at being both. Eggs,
for example, can be delicious in their own right, but are also
exceedingly versatile. Koji belongs in this category as well.

Thanks to koji, we've more or less stopped making long-cooked
meat stocks and producing sauces via classical reductions.
The conventional wisdom about sauces in Europe dictates that
you boil bones—fish bones, beef bones, pork bones, lobster
shells—for hours, then cook down the stock and mount it with
butter. By cooking koji into a lighter broth, we can achieve
the same rich, complex flavors without the heaviness of all that
gelatin and dairy. Koji helps us find the finesse in our raw
ingredients and highlight their natural beauty without smoth-
ering it, like adding a spray of just the right lubricant to
a creaky door rather than a thick layer of grease.

Like a sorcerer's wand, koji transforms other ingredients,
coaxing out both sweet and savory expressions. At Noma, koji
is essential for the production of our pea misos and shoyus.
Additionally, while not essential to the production of meat
garums and fish ferments, we've taken to adding it, not just for
its flavor but also to exploit the enzymes it produces. Adding
koji speeds up fermentation, breaking down proteins and
starches more efficiently. The more you work with it, the more
it becomes an indispensable Swiss Army knife, coming in
handy in all manner of unexpected places.

Oryzae
Loves Grains

Many wild molds are fairly opportunistic, taking hold of
whatever they can lay their spores on, but *Aspergillus oryzae*
is a bit finicky. At Noma, we've tried growing it on a variety
of plants and fruits—from currants to carrots—but few have
worked really well. *A. oryzae* loves grains.

Grains are the seeds of grasses. In the natural world, reproductive success is often a series of trade-offs. Is it better to invest
your time and energy into one offspring at a time or to put
your resources into producing many offspring and hope that
a few survive to maturity? Avocado trees, for instance, take

215

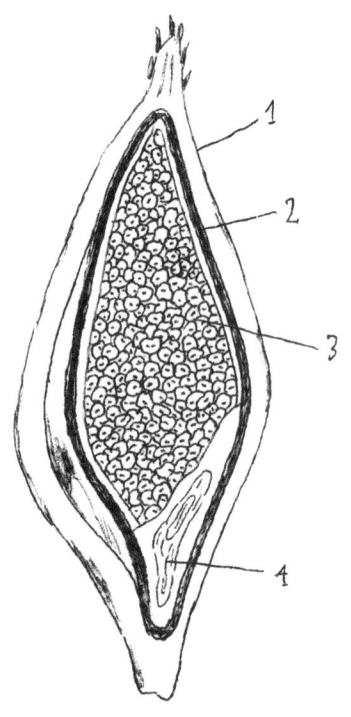

The anatomy of a grain of barley.

1. Hull or husk
2. Bran
3. Starchy endosperm
4. Germ or embryo

Beyond Rice

a long time to reach maturity and bear fruits that also require considerable energy to produce. Their seeds are large and robust—avocado trees pack their kids a nice lunch before sending them into the world. Grasses, on the other hand, are more frugal. They supply the bare minimum to give their offspring a kick start, but they also produce a lot of seeds. The energy they pack for their offspring comes in the form of starch—long, complex chains of linked sugars—packed inside protective husks or shells. Once the seed of a grass like barley germinates, the baby sprout begins producing amylase to break down the starch into the simpler sugar maltose, fueling its metabolism until it can photosynthesize its own food.

The Babylonians and Egyptians, the world's first beer brewers, took note of grain's natural ability to unravel starch into sugar and harnessed it to create the process of malting (from which the disaccharide *maltose* gets its name). Maltsters germinate grain by exposing it to moisture, then stop its life cycle short by roasting and drying the grain. From there, they mix the malt with hot water, after which yeast can ferment the now unlocked sugars into alcohols like beer or mash for whiskey. *A. oryzae* performs the same function as malting. It can work on cooked grains that have never been germinated, uncoiling and splitting the starches within. But where malting focuses solely on the starches in grain, koji also disassembles the nutrient-rich coat of protein that surrounds them.

Protein is made up of amino acids, much like starch is composed of chains of simple sugars. Once unbound, many of these amino acids register on our tongue as the flavor umami. Koji's ability to break down protein (and small amounts of fat) is the key to its extraordinary utility. After all, sweetness counts for a lot, but not everything.

The Italian priest and biologist Pier Antonio Micheli first classified the genus *Aspergillus* in 1729. To him, the shape of the mold's stalk and spores was reminiscent of an aspergillum, the staff used for sprinkling holy water in Catholic ceremonies. The species name, *oryzae*, is derived from the Latin word for "rice." But that tells only a small part of koji's story.

In ancient Japan, rice was the food of the aristocracy. The majority of the population, including peasants and farmers, couldn't afford to eat rice, which they grew and sent to feudal lords as tax payments. Their subsistence diets leaned more toward grains like barley, and as a result, much of the koji in Japan was grown on grains other than rice.

Over time, shifts in the economy and changes in social stratification meant products like miso made with barley koji fell out of fashion. In Japan today, rice koji is the more common and preferred commodity. At Noma, we grow koji on pearl barley.

At first, the decision to use barley was based on locale. We wanted to use this amazing mold, but apply it to our region. Given the rich history of beer production in northern Europe, naturally we turned to barley. Years later, when we moved our entire staff to Japan for a pop-up in 2015, we were excited to try growing koji on rice for the first time, but soon reverted to barley because we found we enjoyed its flavor more. *A. oryzae* reacts differently to being grown on different grains, producing distinct metabolites in response to the quantities of nutrients available. Rice contains more starch than barley does, and as a result, it can get a bit too sweet for our tastes.

Furthermore, when carried through secondary fermentations, as when we lacto-ferment koji for a sour water, the extra sugar fuels fermentation with a bit too much gusto, resulting in off-flavors.

We've experimented with inoculating many other grains, but keep coming back to barley, though that doesn't mean you should ignore other substrates. At Noma, we've inoculated everything from millet to fresh nuts with *A. oryzae*. Rye koji has a certain meatiness and flavors that remind us of Parmigiano cheese. Konini (a heritage variety of purple wheat) inoculated with koji has an intense nuttiness. Wheat koji, a slight breadiness with lots of floral fruit.

All these grains require polishing before being cooked and inoculated. Recall that the starch in a seed is packed within

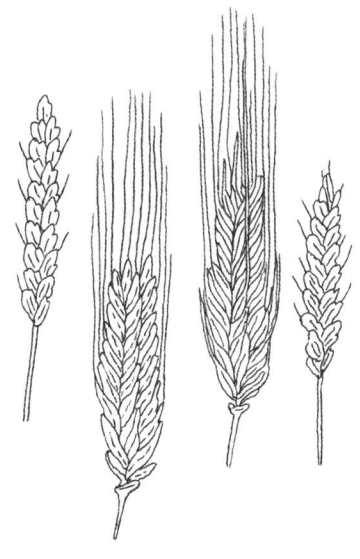

Barley was a foundational crop of ancient civilization and is our preferred substrate upon which to grow koji.

217

husks that are built to be durable and protective. Koji can have a bit of a difficult time getting its hyphae through the outer layers to reach the starch. It's not an impossible feat, but koji grows much better on grains with the husk, bran, and germ removed. You can see this in the grading of Japanese sakes. Sake makers polish off the outer layers of the rice before inoculating it with koji—the more that's polished off, the more expensive the sake. We brought back a countertop grain polisher from Japan specifically to help us get to the inner layers of grains we wanted to turn into koji.

Should you decide to experiment with polishing and fermenting different grains, keep in mind that the proteins in grains sit on the outer region of the kernel, just beneath the bran. Too much polishing can remove this protein coat, which is the source of much of koji's flavor—as well as the umami in some secondary ferments. These flavors are mostly undesirable in sake, but essential to what we're chasing with koji making. In other words, you have to try to polish off the bran while leaving the protein coat intact.

The Many Faces of Koji

Ancient cooks in China and Japan were the first to figure out how to harness and wield koji. More than 2,500 years ago, some bold Chinese culinarian must have decided to taste a batch of moldy cooked grains that had been left out, and discovered it had bright notes of tropical fruit, with sweetness as well as richness. Of course, this was not as safe or simple as it sounds. *A. oryzae* has more than 250 fungal relatives, many of which produce highly carcinogenic poisons called aflatoxins that can be lethal to people with compromised immune systems. But *A. oryzae* is different. Numerous studies have shown koji to be free and clear of aflatoxins and safe for consumption. Still, it's worth noting that *Aspergillus oryzae* is a descendant of a more ancient black mold, domesticated in much the same way that dogs were tamed from wolves. A potentially harmful organism was slowly and selectively made useful and docile over many, many generations.

The first mention of koji in print can be found in *The Rites of Zhou*, a Chinese governmental text from around 300 BCE, which referred to the mold as *qu* (pronounced "chew"). Over the next few centuries, records indicate that qu became a major commodity in China. Instructions for the brewing of grain-based wines and bean pastes appear in official Chinese texts some three hundred years later, and a steady proliferation of knowledge followed. By the eighth century CE, koji had reached Japan.

From there, through random mutations, color variants began cropping up. Albino mutations were selected for and bred. The koji breeders found that they could isolate koji from other wild molds by adding roughly 1 percent ash to their batches of cooked rice or barley. The ash raises the pH of the rice, creating an inhospitable environment for other molds. (*A. oryzae* is tolerant to slightly alkaline environments.) By selectively breeding albino strains, invaders were easily spotted and removed, leaving genetic lines pure. Further mutations led to subspecies that produced different metabolites in varying quantities. The 1,200-year-long breeding process in Japan mirrors the wild bacterial inoculation of cheeses in the caves of *affineurs* in France, where certain cheeses are associated with specific bacterial cultures such as *Penicillium roqueforti* (found in blue-veined cheeses) or *Brevibacterium linens* (responsible for the orange rind of Limburger).

The Japanese used koji to come up with a galaxy of intoxicating and captivating concoctions like miso, shoyu, *amazake*, and sake. Throughout the centuries, the culturing of koji was a closely guarded secret. Fewer than ten koji breeders grew strains of the fungus, selecting for specific qualities spanning generations for close to a thousand years. Miso makers and sake brewers would have to order the spores, known as *koji tane*, in small quantities from these breeders. Eventually, the market opened up, and today Japanese producers grow more than ten thousand specialized cultivars of *Aspergillus*.

Koji is the key to a wealth of fermented foods, from miso to rice vinegar to shoyu to *amazake* and sake.

The myriad variants of *A. oryzae* each have their own qualities and characters. The koji we use for most of our work at Noma is an albino variant of a yellow strain of koji that also goes into producing most sakes. We also use a variety, *Aspergillus luchuensis*, that produces citric acid. It veers away from the overt tropical fruit flavors of *oryzae* and creates hints of green apple and raw oyster mushroom. *Luchuensis* grown on rice produces a straightforward koji, with apple flavors and lemony brightness. But when grown on barley, the mold finds another level of intrigue, generating earthiness and a pleasant bitterness that's almost indistinguishable from grapefruit. *Aspergillus awamori* is an older variety with very black spores. In Okinawa, locals grow it on the *indica* subspecies of rice to produce a distilled alcoholic beverage called *awamori*. Like *A. luchuensis*, *A. awamori* produces citric acid as a metabolite, making for a pleasantly tart koji (though it should be noted that citric acid doesn't distill into the finished alcohol, as it's too heavy to evaporate). At Noma, barley koji grown with *A. awamori* offers us notes of grape must, like saba.

220

You could spend a lifetime investigating the permutations of various molds grown on one substrate versus another. Even then, you'd scarcely scratch the surface of the spectacular diversity of koji.

The main logistical obstacle that faces intrepid koji makers will be sourcing specific spores. Koji's popularity among amateur enthusiasts has yet to break into the mainstream the same way that craft beer has. But just because you don't see koji tane on the shelves of your local grocery store doesn't mean you can't find it.

An Internet search for "koji kin," "koji tane," or "koji spores" will lead you to one of several manufacturers of home-brew sake kits. Many of those companies are located in Japan, although there are a few in North America that have a small selection of different spores. (See page 448 for a list of recommended suppliers.)

Albino strains of *Aspergillus oryzae* have the best flavor, especially for the applications that appear in this book. Varieties like *awamori* and *luchuensis* can be found online with a little digging. If you speak Japanese, or have a friend who does, that will help in figuring out exactly what it is you're purchasing. Generally speaking, however, Latin taxonomic names will be used across the board.

Koji spores are extremely resilient. They ship well and can last for years if they remain vacuum-sealed in the freezer; but even at room temperature, they'll keep on the shelf for up to six months.

A Place for Koji to Call Home

All ferments are complex living things that require environments suited to their needs in order to flourish, and koji is one of the fermentation world's more finicky characters. Its optimal environment is very specific, but don't let that stop you from growing it. Koji matures in less than two days. Even if you mess up the first time (or the first couple of times), you can give it another go without sacrificing too much time or effort. And as with raising children, it gets easier the second time around.

221

A. awamori, grown on barley for 48 hours

Green *A. oryzae*, 48 hours

A. luchuensis, 42 hours

A. luchuensis, 48 hours

But before we start talking about second or third attempts, let's give the process a first look from start to finish.

1. Rinse, soak, and steam the pearl barley until it's fully cooked.

2. Break up the grains and allow them to cool to room temperature, then inoculate them with koji spores.

3. Place the inoculated grains in a fermentation chamber, ideally held at 30°C/86°F and 70% to 75% humidity.

4. Allow the koji to grow for 24 hours. Use gloved hands to turn the grains and furrow them into three rows for better heat dissipation.

5. Give the koji another 18 to 24 hours to finish growing. It will continue to grow, but you want to harvest it before the fungus goes to spore.

The first thing that will probably leap out to you is the need to keep the temperature and humidity at constant levels. The environment you want to create for the koji harkens back to the world *Aspergillus oryzae* arose from: the warm, wet climes of southern China. The fungus also needs oxygen for cellular respiration, so airflow is essential. Airflow is also important because koji produces a fair bit of heat as it grows, and that heat needs to go somewhere. It's not uncommon for a tray of koji in a cramped space to jump to temperatures in excess of 42°C/108°F, at which point the mold will die off.

In Japan, sake makers traditionally grow koji in a cedar-lined room known as a *koji muro*. The shallow trays in which the koji sits are also made from untreated cedar, which has a flavor of its own as well as antimicrobial properties that facilitate the koji's growth by helping to exclude other microbes that might find a bed of cooked grains inviting. The trays are never washed and, over time, *Aspergillus* comes to live in the trays themselves, in much the same way that unique strains of yeasts live in the rafters of the Belgian abbeys famed for their beers.

223

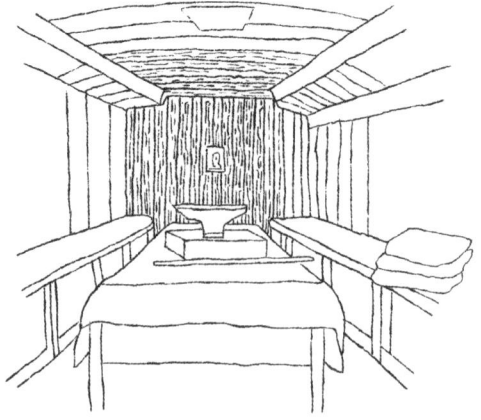

When we started getting serious about fermentation at Noma, we built a series of fermentation rooms inside a few stacked shipping containers. It's almost certainly absurd to ask you to do the same or to convert a room of your house into a nursery for molds, so let's consider some more practical options. There are plenty of solutions, ranging from custom-made wooden cabinets to unplugged mini-fridges to picnic coolers to speed racks with a vinyl cover to laboratory-grade environmental holding chambers. Whether homemade or top of the line, your fermentation chamber needs to be able to do three things: Retain heat. Retain moisture. Allow the koji to breathe.

Decommissioned refrigeration equipment works fantastically well, whether that's an old chest freezer, stand-up cooler, or mini-fridge. These items are all built with insulation in mind, and they can really keep heat and humidity in. Fridges and coolers are also waterproof and easy to clean, which is nice because at such high relative humidity, moisture can collect in the chamber and ruin it, if you're not careful. Coolers or Styrofoam insulation boxes, like the kind you'd take on a picnic, will do the job as well. The one thing you'll need to account for with a fermentation chamber made from a piece of repurposed refrigeration equipment is airflow. You may need to drill a few 1-centimeter / ⅓-inch holes into the top of your fridge or cooler, then cover the holes with a screen or cheesecloth to keep intruders out. Styrofoam boxes can have their lids propped open and left slightly ajar. Alternatively, if your container has an ample amount of space, occasionally opening the door will cycle enough fresh air in.

Continuing with DIY chamber possibilities, wooden cabinetry works fantastically well. It's never so airtight that the koji can't breathe, and it can be cleaned easily. A simple upright rectangular box with a hinged door is all you need. We've also seen ambitious home fermenters who managed to grow koji in dehydrators that have been modified—albeit against the manufacturer's warnings—to have their fans stopped or turned down.

As for the koji-growing trays, it doesn't take a major feat of carpentry to build your own wooden trays from five or six pieces of untreated cedar. Otherwise, perforated stainless steel hotel pans or baking trays will work well when lined with a lightly dampened towel. Perforation is important, though, or moisture will accumulate and drown the mold. Wherever koji can't sprout, something undesirable will.

The size and type of chamber you choose should be informed by the amount of koji you're planning to produce. Cultivating one tray every couple of months won't merit a secondhand soda fridge in your basement, but if you're working in a small restaurant trying to make koji on a weekly basis, investing in a small warming cabinet, proofer, or Winston CVAP (controlled vapor) drawer will make your fermenting life a breeze. CVAPs, digital proofers, and even more advanced combination ovens will all supply you with rack space, along with regulated heat and humidity with the push of just a few buttons. Needless to say, they are expensive pieces of professional equipment that many kitchens would be hard pressed to justify tying up for two straight days.

For DIY solutions, you'll need some means of controlling the heat and humidity in the chamber. For a smaller container like a Styrofoam box, a gentler heat source like a heating pad

226

The schnitzel is a piece of black lip
abalone that's been braised in rice koji
oil to tenderize its flesh, then pounded
and breaded with rice koji flour and
breadcrumbs before being pan-fried.

placed on the bottom of the container will do the job. For a larger chamber like a repurposed fridge, you'll want something like a small space heater equipped with a fan. Our best recommendation would be to purchase a digital temperature controller equipped with a probe and a female power plug that you'll use as the power supply to a separate electric heater. The controller will monitor the temperature inside the chamber and turn the heater on and off accordingly.

As for regulating humidity, strategies can vary from placing a small pot of hot water inside the container to laying a clean damp rag over your koji. The best solutions for small-scale chambers, however, are small ultrasonic humidifiers connected to a humidistat. Neither piece of equipment is prohibitively expensive, and they'll be small enough to rest inside most chambers. As for what 70% to 75% relative humidity looks like in the chamber, you're looking for condensation to just begin pooling at the base of the container. A lightly dampened rag placed in the container shouldn't dry out, but it also shouldn't get any wetter.

For reference, in our fermentation lab, we've built waterproof insulated rooms with electric coil heaters as a heat source. The coils are hooked up to PID (proportional-integral-derivative) controllers, which are computerized thermostats that regulate temperature by way of a feedback algorithm, as they measure the rate of heating with a thermocouple and modulate the power supply accordingly. PID controllers can usually hold temperature steady to within less than 1°C. The humidity is provided by high-pressure nozzles that provide a fine mist according to what a humidistat tells them to do. It's a very elaborate setup. It also works very well. We don't expect you to construct something as elaborate as what we've built at Noma, but it gives you an idea of how to do it well, which is always important when translating things to a small scale.

See page 42 for specific instructions on building a smaller-scale fermentation chamber.

Use a sugar shaker to knock off spores from dried koji grains when inoculating barley.

230

Pearl Barley Koji

Makes 1.1 kilograms

500 grams pearl barley
Koji tane (koji spores; see recipe
for details)

Koji can be finicky. *Aspergillus oryzae* demands specific conditions to grow. You'll have to go to a little bit of trouble to create an environment that's conducive to its survival.

With that being said, koji is a robust mold that *wants* to live and succeed. Other molds, like red yeast rice, can take up to 7 days to grow before they're ready to harvest. Koji matures in less than 2 days in optimal conditions. Those conditions are very specific, but very achievable, even at home.

Equipment Notes

You'll need to construct a fermentation chamber, which you can read about in the preceding section or see "Building a Fermentation Chamber" (page 42) in the primer chapter. You'll also need cedar or perforated plastic or metal trays that fit into your chamber. If you choose to use cedar, be sure to use untreated wood. A perforated half-gastro/hotel pan (32 × 26 centimeters / 12½ × 10 inches) is ideal for this amount of koji. Any smaller, and there won't be enough room for air to circulate around the grains. It can help to grow koji on clean cotton kitchen towels that will soak up excess moisture. And finally, as with handling many sensitive microbes, wearing latex or nitrile gloves will help keep things sanitary.

One final note: This recipe will work just as well with *japonica* (short-grain) rice, if you'd like to make a more traditional koji.

231

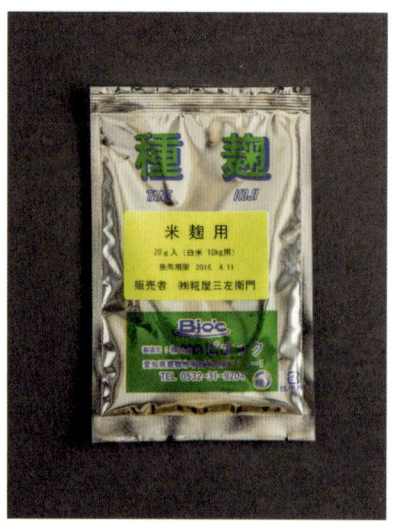

A packet of *koji tane* from Japan.

In-Depth Instructions

Koji tane is available as two different products: powdered spores and dried grains of rice or barley that are covered in spores. They can be purchased in various package sizes online and from home-brewing shops (see Sources, page 448). A little bit goes a long way. A 100-gram bag contains enough spores to inoculate 100 kilograms of grains. It's also an investment you only really need to make once, because once you've made your own koji, you can grow your own spores from it for future use (see "Harvesting Your Own Spores," page 241). Both versions of koji tane can be used to inoculate fresh koji. (Don't worry about inoculating barley with spores grown on rice, or vice versa; koji's an equal opportunity consumer.)

To begin, place the barley in a large bowl and fill it with cold water. Agitate the grains with your hand, then pour off the cloudy water. Repeat once more, then fill up the bowl with water for a third time, bringing the water line above the barley by a few centimeters. Allow the barley to soak for at least 4 hours at room temperature, or overnight in the fridge.

Once the barley has soaked, drain and rinse the grains in a colander until the water runs clear. Shake off any excess.

The ideal medium for *Aspergillus*'s growth is on hydrated but fully separate and relatively plump grains. Boiling can easily overhydrate them, leaving them wet and mushy. The mold can grow too quickly in this situation, reaching its reproductive cycle before producing the concentration of enzymes we're seeking. If the grains are very wet, the spores will effectively drown and never start growing. Steaming allows the grains to fully cook without taking on excess water. At Noma, we have combination ovens that allow you to add steam to a convection cooking cycle—they're brilliant for steaming grains—but a classic steamer will also work just fine, as will a sieve or colander set into a pot with a lid.

If using a combination oven: Steam the barley at 100°C/212°F for 45 minutes with the fan speed set to 80%.

If using a regular steamer: Cook the grains for about 20 minutes over simmering water, but begin testing at 15 minutes. Bite into a kernel—a properly cooked one should still be dense but easy to chew and should not be hard or white at its center.

While the barley is cooking, ready the tray in which you'll be fermenting the koji. If using cedar, make sure it's clean and free of debris; if it's metal or plastic, make sure it's washed and sanitized. Line the metal or plastic tray with a towel. If it's a kitchen towel, make sure it's clean (and was washed without perfumed detergent), sanitized with steam, and wrung dry.

Once the barley is steamed, you want to break it up while it's still warm so that any residual starch doesn't set and clump the grains together. Wearing two pairs of latex or nitrile gloves will protect your hands from the heat. Rub the grains between your hands into your fermenting tray. Don't apply so much pressure that it ruptures the grains; keeping them whole is ideal for koji's growth. Spread the grains out and allow them to cool to 30°C/86°F on a cooling rack on the counter. Feel free to fan the grains if you're impatient (although if you're impatient, you probably shouldn't be growing koji). With the grains steamed, separated, and cooled, it's time to inoculate.

If using powdered spores: Put a small amount of the powder into a tea strainer and gently tap it over the grains. The spores are extremely potent—1 teaspoon contains more than a billion spores, plenty to inoculate one tray of barley. Wearing gloves, fold the barley thoroughly, being sure to get into the corners, and then give it another pass with the spores. Turn the grains once more, and you're set.

If using koji tane on dried grains: Fill a steel shaker (like you'd use for powdered sugar) halfway with the koji grains and make three passes rather than just two, as there will be fewer total spores delivered to the grains. Use gloved hands to fold the barley over between each pass, and make sure to reload the shaker with more koji if you don't see a cloud of spores coming out of the shaker.

It's important to soak barley until well hydrated before cooking.

233

Barley Koji, immediately after inoculation

After 30 hours

Spread the inoculated barley in an even layer, pulling it in from the sides of the tray to avoid creating any spots without access to airflow. Cover with a clean, barely damp kitchen towel, making sure there are no exposed areas.

Transfer the tray to your fermentation chamber. Raise the tray off the bottom of the chamber with a trivet or wire rack to allow for airflow all around the koji. In any chamber that's smaller than a broom closet, keep the lid or door open just a crack to allow fresh oxygen in and excess heat to escape. We saw the mold suffocate all too many times in the early days of our koji-growing at the restaurant. Insert a thermometer probe into the bed of barley, and be sure your humidity meter is running.

No matter what setup you're using, keeping tabs on the temperature of the grains is paramount to your koji's success. At room temperature, the koji's growth will be severely retarded and the organisms will struggle to take root. But any hotter than 42°C/108°F and you'll cook them to death. Humidity also has to be monitored carefully. It's easy to drown the koji if the environment gets too wet. On the other hand, should the grains dry out, the hyphae will meet too much resistance and fail to penetrate the starch. Insert the temperature probe into the grains and set the temperature controller for 30°C/86°F. Keep the humidity between 70% and 75% for the duration of the ferment. Refer to the instructions on building a fermentation chamber (page 42) for tips on how to achieve this.

Assuming all goes well, after 24 hours you should notice the first inklings of mold growth: wispy, ill-defined white threads should begin to cover the barley, lightly binding the grains together. Remove the koji from the incubator and set it on a clean rack on a countertop. Again wearing gloves, break up the grains in the same manner as when they were first cooked. Mixing the grains brings koji growing at the bottom of the tray into contact with air and breaks up the mycelium, encouraging further growth as it tries to spread its web.

After you've turned the grains and broken up any clumps, furrow the grains into three rows, like mounds in a farmer's field. This creates more surface area for contact with fresh air and improved heat dissipation.

After the first 24 hours of growth, koji's metabolism will kick into overdrive. Your job is to keep it alive for the next 24 hours. Return the koji to the fermentation chamber, with the thermometer probe in the center of the middle mound. If you notice the heat spiking, adjust the temperature controller, and open the door or lid for 30 minutes to allow the chamber to cool down. If you're worried the koji is still on the path to overheating, breaking it up again will help to lower the temperature.

Over the next 12 hours, the koji's mycelium will firmly bind the grains of barley together, forming a dense cake. By hour 36, the koji will be covered in a light green or white fuzz (depending on the strain of *Aspergillus* you're using), but full enzyme production and flavor won't develop until hours 44 to 48. At that point, the smell of the koji should be intensely fruity, like ripe apricot.

To harvest your koji, you need to halt its growth. Place the entire tray in the fridge for 12 hours to cool down. You can then pack it into airtight containers to store in the fridge for a couple of days if you have plans to use it in the near future. We actually find that koji's flavor improves considerably after a couple of days in the fridge. It's perfectly ready to use as soon as you harvest it, but the refrigerator will arrest the koji's growth while the enzymes continue to work, sweetening the koji even more in the process. If you don't have plans to use it immediately, koji stores well in airtight containers in the freezer for up to 3 months.

To complete the koji-making process, wash your equipment—including the fermentation chamber, the trays (if not made from wood), and probes—to keep things sanitary. If you're using wooden trays, simply wipe them clean with a lightly dampened towel and let them air-dry in a well-ventilated area.

After 42 hours

After 48 hours

237

1. Barley and koji spores (*koji tane*).

2. Rinse the barley thoroughly under cold water, then soak for 4 hours.

3. Steam the barley until tender but not falling apart, 20 to 30 minutes.

4. Break up the barley while still warm to avoid clumping.

5. Let the barley cool to at least 30°C/86°F.

6. Inoculate the cooled barley with koji spores.

7. After 24 hours, you should begin to see the signs of mycelial growth. Mix the barley and furrow into three mounds.

8. After 48 hours, the koji should be fully grown.

9. Cool the barley in the refrigerator to stop the fungus's growth. Pack in airtight containers and store in the refrigerator or freezer.

If you're making koji on a regular basis, at some point you may want to harvest spores.

Make a batch of koji and use gloved hands to break up the grains onto a sanitized, nonreactive, nonperforated tray. The mycelium will be very strong by hour 48, so you may have to pry the grains apart. Do your best to avoid crushing the barley. Spread the grains out in a single layer to create the maximum amount of surface area for the koji to sprout its spores. Cover with a lightly dampened towel and return the koji to the fermentation chamber. Allow it to grow for another 36 hours. Continue to monitor the temperature and humidity, but you won't need to do any more turning of the grains.

After 36 hours (84 total), you should see fluffy white, green, or yellow spores, depending on which strain of *Aspergillus* you purchased. If you touch the grains (wearing sterile gloves), your finger should come away covered in powder. If agitated, the grains will emit a plume of dusty spores with a strong, almost meaty scent. If grown properly, the spores should be plentiful.

You'll need to dry the grains to keep the spores shelf-stable and to prevent them from becoming infected by other microbes. Remove the towel that was covering the koji and the humidity source from your chamber, and dry off any moisture from the walls. Crack the lid or door a little wider than when you were growing the koji, in order to increase airflow and place the sporulated grains in the chamber to dry until completely hard, about 2 days.

Pack the dried koji into an airtight container and store in a dark cupboard for up to 6 months or freeze for longer-term storage.

Suggested Uses

Crunchy Koji Croutons

You can do a lot with your fresh cake of koji, but it's also delicious on its own, cut into slabs and seared in hot oil. If you want to take things a step further, you can cut the cake into bite-size cubes and deep-fry them into crisp, golden croutons. Drain them on a paper towel and season with salt. They have a sweetness and deep umami that combine to leave you thinking you've just had a delicious bite of ham fat. Double down on this effect by draping a thin slice of Ibérico ham over your koji croutons for an addictive bite-size open-faced sandwich. Or brush the fried koji squares with Beef Garum (page 373) for additional saltiness that balances out the natural sweetness of the koji cake.

Stews and Soups

Crumble fresh koji cake into pieces the size of a kidney or fava bean and add them to vegetable soups during the last 10 minutes of cooking. Use as much as you would of any other vegetable in your soup, to lend a note of sweetness and a surprising texture that's reminiscent of a chewy dumpling. When making a meat stew, you can add the crumbed koji in the final hour; it will thicken the broth but will also add an extra element of sweetness and richness to the whole dish.

When inoculated with *Aspergillus luchuensis*—
a different species of *Aspergillus*—barley koji develops
notes of citrus.

Citric
Barley Koji

Makes 1.1 kilograms

500 grams pearl barley
Albino *Aspergillus luchuensis* spores
  (see Sources, page 448)

Aspergillus luchuensis is harder to procure than its mainstream counterpart, *A. oryzae*, but its flavor is remarkable, reminiscent of green apple and lemons. We were thrilled at Noma when we came across this special mold, because the citric acid it produces brings a sharp layer of contrast to the umami-rich flavor of koji. It's worth seeking out. Just be warned that *A. luchuensis* has a melanistic strain that is pitch-black with quite different flavors; you want the albino version. Inoculating barley with *A. luchuensis* is functionally the same as using *A. oryzae*, but there are some finer points that will need your attention.

We recommend that you first read the in-depth instructions for Pearl Barley Koji (page 231), as it serves as a template for this recipe.

Rinse, soak, and steam the barley as directed for Pearl Barley Koji. Separate the grains onto a perforated tray lined with a clean towel. If using *A. luchuensis* spores that are still attached to their grains, inoculate the barley using a shaker; if using powdered spores, use a tea strainer. But know that the spores of *A. luchuensis* are far more copious than those of *A. oryzae*. Two passes with the shaker is more than sufficient, and it's unlikely that you'll have to replenish the grains between passes. If you're using powdered spores, 1 teaspoon should be enough to inoculate a full tray.

243

Citric Barley Koji, immediately after inoculation

After 30 hours

After 36 hours

Transfer the koji tray to the fermentation chamber. Citric koji likes things a little cooler than *A. oryzae*, so aim to keep the temperature at 28°C/82°F rather than 30°C/86°F. The humidity is fine at 70% to 75%.

After 24 hours, use gloved hands to turn and furrow the grains into three rows. At this point you should definitely be able to taste a pronounced sweetness and the recognizable twang of citric acid. Over the next 24 hours, the sweetness and acidity will build steadily, bringing out other profiles in their wake. At hour 36, the citric koji will have a not-so-subtle taste of lemon, green apple, and oyster mushroom. This is when you want to harvest it, as opposed to letting it run for the full 48 hours that you would for koji inoculated with *A. oryzae*. Left to its own devices, *A. luchuensis* will break down the barley much more thoroughly than its cousin, and a very discernible bitterness begins to come through after hour 40 that makes the koji a dead ringer for grapefruit. Beyond that, the bitterness intensifies while the sweetness dissipates. (Interestingly, the bitterness is nonexistent when *A. luchuensis* is grown on rice, though its overall flavor is less exciting.)

To harvest the koji, place the entire tray in the fridge for 12 hours to cool down. You can then pack it into airtight containers to store in the fridge for a couple of days if you have plans to use it in the near future, or freeze it for use further down the road. Koji will keep for up to 3 months in the freezer.

To harvest *A. luchuensis* spores, spread the grains into a single layer and continue to incubate them for another 36 hours. Dry the sporulated koji according to the instructions on page 241 and pack into a resealable bag or airtight containers for storage.

After 42 hours

245

Sweet Citric Koji Water

Makes about 1 liter

1 batch Citric Barley Koji (page 243),
     harvested at hour 42

In this recipe, we extract the tart flavors produced by our citric koji into a liquid that can be used almost anywhere you'd ordinarily use white wine for cooking. Amazingly, almost none of the bitterness of 42-hour-fermented koji transfers to the clarified liquid, which itself tastes like a mixture of apple juice and lemon tea, with a subtle hint of earthiness.

For this recipe, you'll make a batch of Citric Barley Koji, but let it grow for a few additional hours. The citric acid content, which is the key, is noticeably higher at hour 42 than hour 36.

Measure out twice the koji's weight in water (about 2.2 kilograms) and blend the two together in a blender at high speed for 1 minute; work in multiple batches, if necessary. Transfer the blended mixture to a freezer-safe container with a lid or heavy-duty zip-top bags, leaving enough room for the mixture to expand as it solidifies. Place in the freezer.

Once the koji mixture is frozen solid, transfer it to a colander or sieve lined with cheesecloth and set it over a deep container to catch the liquid as it thaws. Cover the colander with a lid, transfer to the refrigerator, and leave it to thaw for 3 to 4 days. Once all the ice has melted, carefully remove the colander, discard the solids, and harvest the clear yellow liquid.

Store, covered, in the fridge for up to 5 days, or in the freezer if you're not planning to use it immediately.

Suggested Use

White Wine Stand-In

Think of sweet citric koji water as white wine with umami and body. Use it to steam cockles or to liven meat jus. It's also perfect for making fish *en papillote*: Add citric koji water to your parchment parcel in place of wine, along with citrusy herbs like lemon thyme or pineapple sage and light vegetables like baby summer squash. Bake in the oven for about 15 minutes, then peel open the paper to reveal perfectly cooked fish infused with the sweet tartness of the koji water.

247

Citric koji amazake doubles as a beverage
and a cooking medium.

248

Sparkling Citric Koji Amazake

Makes about 2 liters

2 kilograms (1 batch) Citric Barley
     Koji (page 243)
2 kilograms water
1 packet (40 milliliters)
     Chardonnay yeast
½ Campden tablet (optional)

Amazake is an iconic sweet Japanese beverage made from rice koji, rice, and water. It's sometimes fermented and mildly alcoholic, sometimes not, sometimes pureed or strained, and other times left chunky. In most recipes, freshly cooked rice is mixed with equal parts rice koji and water and then held in a rice cooker on "keep warm" for 6 to 8 hours. The temperature is the key to making amazake, as the enzymes the fungus produces are highly effective at 60°C/140°F, turning starches to sugar as they hop from one catalytic reaction to another.

At Noma, our greatest success in this realm has been a very unorthodox version of amazake made of barley fermented with *A. luchuensis*. (Though if all you can find is garden variety *A. oryzae*, don't let that stop you.) The end result is a little alcoholic, though it's not beer and it's not sake. Truth be told, we've never actually served it at the restaurant, but it's so damn delicious that we had to include it.

Combine the koji and water in a vacuum bag, seal it, and cook it in a circulating water bath set to 60°C/140°F for 8 hours. Alternatively, you can place the mixture directly in a rice cooker set to "keep warm" or into a fermentation chamber.

Strain the liquid through a fine-mesh sieve lined with cheese-cloth, squeezing out as much as you can without forcing any solid matter through the sieve. Discard the mash.

Chardonnay yeast ferments the sugar in citric koji into alcohol.

An airlock will allow the amazake to vent as it ferments.

Allow the amazake to cool to room temperature, then stir in the Chardonnay yeast. Transfer the mixture to a fermentation bucket, carboy, or glass jar and cap it with the airlock. (This is all standard brewing equipment that can be procured at any home-brew shop.) Allow the mixture to ferment in a cool basement or garage for 4 to 5 days. You're looking for a mildly alcoholic beverage, in the realm of a light cider or beer. It will be lightly effervescent and a fair bit dryer than what you started with.

As with kombucha, we need to stop the fermentation well short of how far it *could* go. Amazake has a lot of fermentable sugar and we want to keep much of it in the finished product. Bottle and drink the amazake soon after it's brewed. Chilling will bring the fermentation to a crawl, but as a living ferment, its sugar content will continue to decline even at fridge temperature.

If you want to halt fermentation completely, one alternative is to stir in a "sterilizer" like a Campden tablet (made from potassium metabisulphite), which will halt the yeast's ability to reproduce. They're readily available at home-brew shops and online. Half a tablet should suffice for the amount of amazake this recipe yields. The only other option is to sterilize the amazake in a hot-water bath: Fill a few swing-top bottles 85 percent full, cap them, and place the bottles in a pot of water at 70°C/158°F for 15 to 20 minutes. This is effective at killing the yeast, but it also kills some of the flavor. There is no real substitute for drinking this fresh.

Suggested Use

Clams and Cockles

The amazake has qualities that ride the line between a beer, a cider, and a young wine. You can and should absolutely cook with it, adding a dash to soups or stews to perk them up, or using it wherever you might otherwise reach for a bottle of white wine. The amazake complements bivalves extremely well, for instance. Steam a kilo's worth of clams or cockles with olive oil, a clove of garlic, a couple of sliced shallots, and a few glugs of amazake. Once the bivalves have opened, remove them from the pot and return the remaining liquor to the stove. Whisk in a knob of butter, add some chopped tarragon, parsley, and chives, then pour it back over the clams before serving.

Citric koji amazake is ideal for steaming seafood.

Above: Sift dried, milled koji to make koji flour.
Opposite: Dry crumbled koji in a dehydrator
set to 50°C/122°F.

252

Dried Koji
and Koji Flour

Makes about 500 grams

1 kilogram koji of any variety

Drying koji completely transforms its usability as an ingredient in the kitchen. In your dry pantry, you end up with a completely new ace in the hole that occupies a space between an exotic sugar and all-purpose flour.

Use your fingers to break up the koji as finely as possible, and spread it onto a tray lined with parchment paper. Set your dehydrator to 50°C/122°F and dry the koji until it's completely desiccated—usually 24 hours. At this point, you can pack the dried koji into airtight containers and store it in the freezer for up to a few months.

Otherwise, blend the dried koji into a fine flour in a blender on high speed—45 seconds to 1 minute should suffice. To ensure that no coarse grains remain, sift the powder through a fine-mesh sieve or tamis over a large bowl, circulating your hand through the powder to drive all the flour through. Any dried koji that doesn't pass through the sieve can be reblended and sifted again. The koji flour is full of sugar and is therefore fairly hygroscopic (moisture-loving), so be sure to store it in an airtight container at room temperature.

253

Boil koji in water for koji stock.

Suggested Uses

Koji Stock

One of the best possible uses for dried koji is as a flavoring for stock. Boil 1 liter water in a pot and add 150 grams crumbled dried koji (not koji flour). Turn down the heat and let the stock simmer for 10 minutes. Strain and discard the solids. What you have is a versatile, vegetarian base liquid that can be used for a whole flight of applications.

Koji-Miso Soup

Once you've made koji stock, you can use a handheld blender to incorporate the stock with any of the misos from this book for a miso soup unlike any you've had before. A rich miso soup will have somewhere in the neighborhood of 20 percent miso (200 grams for 1 liter of koji stock), but all misos differ in their saltiness and intensity. Remember that you can always add more miso, but you can't take it out. Try starting with, say, 100 grams of Yellow Peaso (page 289) and working your way up from there.

Koji-Blanched Vegetables and Koji Soup

Koji stock is great for blanching vegetables—something we practice often at Noma. Picture a roast bird for dinner, with a platter on the side of young carrots, turnips, and cabbage leaves, each blanched in umami-rich koji stock, seasoned with salt, and drizzled with olive oil. Once you've finished dinner, add the bird carcass and leftover pan juices to the blanching pot. Bring the stock back up to a boil, drop the heat, and let it simmer for a few hours before straining. Finish with a splash of vinegar and shoyu for an unbelievably satisfying soup. If you like, you can boil a few peeled potatoes in the broth, then use a handheld blender to make creamy potato soup that can be topped with cooked nettles or spinach. And even if you omit the bird entirely, you can still repurpose the blanching liquid as a delicious vegetarian soup.

Koji oil is subtly sweet and fruity, and will help break down tough proteins if used as a cooking medium.

Koji oil makes an exceptionally savory mayonnaise.

Koji-Infused Oil

Choose an oil that doesn't taste like much—grapeseed, sunflower seed, canola, or corn oil will all work. Combine 250 grams dried koji (not koji flour) and 500 grams oil in a blender and blend at high speed for 6 minutes, until you have a smooth, silky liquid that almost has the texture of cream. Transfer to a container and cover, allowing the oil to infuse and settle in the fridge for 24 hours. The following day, strain the oil through cheesecloth or a fine-mesh sieve and discard the sediment.

In koji oil, you have an ingredient that can be combined with cucumber water and a squirt of lime juice or herbed Celery Vinegar (page 187) for a sublime dressing for thinly sliced fresh scallops. (To make cucumber water, blend an English cucumber to a pulp, then squeeze the mash through a clean kitchen towel to harvest the liquid in a bowl.)

Koji Oil Confit

Slow cooking in koji oil also lends itself well to a confit. When blended into oil, the enzymes in koji act as a great tenderizer for tougher cuts of protein, from abalone to goose legs.

Koji Mayonnaise

Or use koji oil to make mayonnaise. Most people don't make their own mayonnaise, but they should. The difference in flavor and texture is worth the reasonable amount of effort, especially if you use koji oil. Whisk together 2 egg yolks, 1 teaspoon Dijon mustard, and a splash of vinegar. Slowly incorporate about 150 milliliters koji oil in a thin, steady stream, whisking the whole time to emulsify the mixture into a thick mayonnaise. Finish by seasoning with salt and cracked pepper. No sandwich will ever be the same.

Koji-Breaded Fish

Koji flour can step in as a replacement for regular flour in a number of instances. The next time you bread a veal cutlet for schnitzel, dredge the meat in koji flour, then through an egg wash, and then in your preferred breadcrumbs. Koji flour brings a sweet nuttiness to the finished product that regular flour just doesn't have. It works equally well for breaded fish. If you're pan-frying a thin fillet of something like sole, flounder, or halibut, simply dredge it in koji flour before dropping it into a pan of foaming butter. One thing to watch out for is that koji flour caramelizes faster than regular flour. Keep the heat just above medium so as not to scorch the crust before the fish is cooked.

Koji "Marzipan"

Koji flour can be made into a marzipan-ish product by whisking together equal parts by weight of neutral oil and koji flour, then adding 10% of the total weight in powdered sugar (i.e., 100 grams koji flour, 100 grams grapeseed oil, and 20 grams powdered sugar). You end up with a paste that can be used in just about any situation where you'd look to use marzipan, from croissants to layer cakes. Or simply crumble up the koji marzipan over the top of vanilla ice cream to create a kind of next-level cookie-dough sundae.

Bread fish in koji flour before pan-frying.

257

Lacto-fermentation takes koji to an entirely different
(and sour) place.

Lacto Koji
Water

Makes about 1.5 liters

750 grams Pearl Barley Koji (page 231)
1.5 kilograms water
45 grams non-iodized salt

By lacto-fermenting koji, we get a fascinating sweet-sour-savory liquid. It's hard to pinpoint all the facets of its flavor, but we use it everywhere in the restaurant, from the juice menu to sauces, marinades, pastes, and more. Should you make a large batch, keep small bags of it in the freezer so you can always have a little secret flavor weapon on reserve.

Blend all the ingredients at high speed into a puree, about 45 seconds. Work in multiple batches, if necessary, but be sure that every batch has an even distribution of the ingredients.

Seal the mixture in a large vacuum bag (or multiple bags, if necessary), forcing out as much air as possible without spilling the contents. You can also use a large zip-top bag and squeeze all the air out by slowly lowering it into a tub of water, stopping a few centimeters from the top. The pressure of the water will force the air out. Seal it shut and you'll have an effective, albeit imperfect, vacuum.

Let the koji water sit at room temperature or just above for 5 to 6 days. As the mixture ferments, it will produce carbon dioxide and inflate the bag, meaning you may need to "burp" it to prevent it from bursting. Use the same method as with other lacto-ferments: Snip a corner of the bag, force out any gas, and reseal the bag.

259

After being blended, frozen, and thawed, lacto-fermented koji yields a clear amber liquid.

Each time you burp the koji water, be sure to taste it with a clean spoon. As the mixture ages, sweetness will diminish as lactic acid builds. You're chasing a balance where the koji water bites your tongue sharply, but still has some residual sweetness.

Once it's done fermenting, cut open the bag and transfer the liquid to a freezer-safe container with a lid, leaving room for the liquid to expand as it solidifies—about a finger's width from the top should do it. Place the koji water in the freezer.

Once the water is frozen solid, transfer the brick to a colander lined with cheesecloth and set it over a deep container to catch the liquid as it thaws. Cover the colander with a lid and transfer it to the fridge, allowing the lacto koji water to thaw over the course of 3 to 4 days. Once all the ice is melted, carefully remove the colander, discard the solids, and harvest the liquid beneath.

The koji water is very much still a living ferment and can continue to change in all manner of directions if not stored properly. Freezing it (in bags or mason jars) is the best way to stabilize its flavor, though it will be fine for a few days, covered, in the fridge. Alternatively, if you really want to extend the shelf life, you can sterilize the koji water in a hot-water bath: Fill a few swing-top bottles 85 percent full, cap them, and place the bottles in a pot of water at 70°C/158°F for 15 to 20 minutes. This will kill the bacteria and allow you to keep the lactic koji water for much longer. That being said, the flavor is at its purest when untreated and consumed quickly after harvesting.

Suggested Use

Lacto Koji Butter Sauce

Lacto koji water produces an unearthly butter sauce. In a saucepan over medium heat, warm 2 parts (by weight) lacto koji water to a bare simmer, then emulsify with 1 part cubed room-temperature butter by whisking it in one cube at a time or blending it with a handheld blender. Season with salt and reserve in a warm place until ready to serve with roast chicken or fish, root vegetables, or cooked grains. To amp up the pleasure, grate white or black truffles over a slowly cooked creamy omelet, and flood the plate with koji butter sauce.

Koji butter sauce is also an incredible cooking medium. Use it to poach anything from lobster tails to turnips, or add a splash to a pot of wilted kale or a pan of gnocchi.

Lacto koji butter sauce is a revelation—rich and full of umami.

261

Above: Koji mole is a blend of cream and dark-roasted koji.
Opposite: Pass koji mole through a tamis for a velvety sauce.

262

Roasted Koji "Mole"

Makes about 1.5 liters

500 grams Pearl Barley Koji (page 231)
500 grams heavy cream
500 grams whole milk

While this isn't a mole in any traditional sense, we think of it that way. It has depth and complexity and light sweetness like our favorite genuine moles. Adding a splash of vinegar and some ground chile makes it even more like its namesake. Alternatively, you can add a few spoonfuls of this koji mole to a braise for richness and umami as well as a bit of body.

Use your fingers to break the koji into small pieces and spread them on a baking sheet. Roast in the oven at 160°C/320°F, turning and shaking the koji every 10 minutes to make sure it cooks evenly. After 45 to 60 minutes, the koji should smell like roasted coffee and be deep brown in color. Remove the baking sheet from the oven and allow to cool to room temperature.

Weigh out 375 grams of the cooled roasted koji (the original weight will have changed with roasting) and place it in a sealable container. Pour the cream over the top and leave it in the fridge to soak overnight.

Transfer the mixture to a blender, add the milk, and blend everything into a smooth puree. It should take about 6 minutes of blending (if it has trouble catching, just add a little more milk until it spins). If you want to refine the texture even more, pass the mixture through a tamis while it's still warm from the blender. Store in an airtight container in the fridge for up to 4 days, or freeze for up to 6 months.

263

New potatoes glazed in roasted koji mole.

Suggested Uses

Koji Mole–Glazed Potatoes

Koji mole is a fantastic dressing for boiled new potatoes or fingerlings. Place a couple of handfuls of new potatoes in a pot of cold, salted water and bring to a boil. Reduce the heat and simmer until tender, then drain and return the potatoes to the pot. Kill the heat on the stove and, while the potatoes are still hot, add a couple of spoonfuls of koji mole. Season with salt and, if you can afford it, serve with a couple of spoonfuls of caviar or trout roe.

Not Chocolate

Roasted koji turns into a mole analogue so well because it is reminiscent of chocolate. As such, we've found it makes a very distinct yet familiar version of hot chocolate. Blend 60 grams koji mole into 500 grams milk, along with 15 grams muscovado sugar. Warm it up and enjoy on a chilly winter's day.

Koji Cure
(Shio Koji)

Makes about 800 grams

400 grams koji of any variety
400 grams water
40 grams sea salt

Shio koji is a mixture of salt and koji used widely in Japan as a cure for meats and fish as well as a condiment. As a cure, it simultaneously seasons and tenderizes, as the protease produced by koji breaks down animal proteins.

Whir the koji, water, and salt together in a blender. You don't need a smooth paste, just a uniform mixture. If you want to reap the enzymatic benefits of koji, use it immediately as a marinade (see below). But if you leave it longer, the mixture effectively ferments itself, offering up deeper flavors, while the salt takes a step back. The relatively high salt content will allow it to keep well, covered, in the fridge for weeks.

Suggested Uses

Marinade

Shio koji is classically used as a marinade. It improves the texture and flavor of meats that often need a little extra help to reach their taste potential by tenderizing, seasoning, and imparting umami and floral sweetness. We find it works extremely well with game birds, but no less so with the common everyday chicken. For a 1-kilogram / 2¼-pound chicken, slather all of the bird's skin with a thin layer of koji cure and let it sit at room temperature for about 3 hours before roasting. If the bird is smaller—say, a 500-gram / 1-pound Cornish

265

Shio koji is a paste of water, salt, and barley koji.

hen—then cut the curing time in half. A larger bird, such as a duck, can easily take 4½ hours to cure. Turkey, pork, and flank steak also all benefit greatly from a shio koji rub. Meatier varieties of fish like monkfish, pike-perch, or black cod all fare well with a little time in the cure, which firms up texture and drives seasoning and flavor into the flesh. Just be careful: Fish are more delicate than birds or red meat and thus more diligence is needed when curing. Pay attention to the details: How thick is the fillet? Does it taper? If so, apply less shio koji to the thinner sections. A good, chunky 160-gram portion of black cod can spend as little as 30 minutes covered in a thin layer of cure before it becomes too salty. For thinner fillets, aim to cut that time to 15 or 20 minutes.

Finally, before cooking anything you cure in shio koji, be sure to remove as much of it as possible. A spoon or the back of a butter knife is a useful tool for scraping off the cure, followed by a gentle wipe with a paper towel.

Shio Koji Butter

Some people enjoy shio koji as a straight-up spread on bread or as a primer for avocado toast. If the taste is a little too strong for you, try mixing a teaspoon of shio koji into a couple of tablespoons of softened butter to form a compound butter that can top grilled corn or baked potatoes, or be used to finish porridges or stews.

Cure a Cornish hen in shio koji to season
and tenderize it at the same time.

Misos and Peaso

—

Yellow Peaso 289

Rose Peaso 302

Ryeso 307

Maizo 312

Masa 315

Hazelnut Miso 317

Breadso 321

Pumpkin Seed Miso 325

Broadening
Our Horizons

Noma is a restaurant that has evolved in stages. In our earliest stage, when we had barely opened our doors, we were exploring ingredients and becoming familiar with the particulars of each season, looking for nuggets of inspiration from which we could create a dish or a menu or an identity. As we progressed and became comfortable in our skin, we found ourselves with more time and desire to learn not only about ingredients, but also technique, history, stories, and people.

We began to dig deep into the centuries-old foodways of the Nordic region in search of pillars around which we could define our own cooking. But as we sifted through the traditions of Scandinavia, Finland, Greenland, the Faroe Islands, and northern Germany, the cuisine we were seeking remained frustratingly elusive. In retrospect, it makes sense. Our staff comes from a large and diverse set of backgrounds— to restrict our gaze to the Nordic region was never going to be enough.

And so, sometime in 2009 or 2010, in order to learn more about ourselves, we began looking further afield. At first we thought we should focus on countries that were close to ours, like Russia or Germany, where there are a great many shared food traditions. But again, it quickly became apparent that we needed to look outside our comfort zone.

A few trips to Japan opened our eyes to the incredible rigor with which Japanese cooks and artisans pursue umami—the hard-to-describe yet deeply recognizable flavor of savoriness. There are thousands of varieties of shoyu, koji, and miso, each with different characteristics and applications. As a result, chefs in Japan have multitudes of seasoning tools at their disposal, all of which taste distinct but also distinctly Japanese. We returned to Copenhagen knowing that in order to construct a flavor that could define our restaurant and possibly our entire region, we needed a similar pantry of our own.

Our Nordic version of miso is perhaps the most successful attempt we've had at building such a pantry. And it began its life as a failed tofu.

Koji, legumes, and salt go into a bucket and emerge months later as miso.

Soy Story

We were trying to see if we could coagulate milk squeezed from local yellow peas into a sort of tofu. We committed weeks to the effort before realizing that this high-protein legume might actually be better suited for fermentation. From there, we followed traditional miso-making techniques—replacing Japanese inputs with Nordic ones—and ended up with yellow pea miso, aka peaso.

Peaso doesn't taste like miso—at least we don't think so. It tastes distinctly of Denmark, making it both uniquely ours and still greatly indebted to Asia. It's a perfect collision of a concept that is foreign to this part of the world with an ingredient with which we're very well acquainted. That's how we think food should move forward. Most of the best things in Scandinavia—and the world, for that matter—have a similar story: They get introduced to a new place, they adapt, find their own life, and eventually become of the place. It's like immigration. Microbial immigration.

Since that initial success, we've made miso from rye bread and corn and hazelnuts. We've tried different legumes, from Scandinavia and beyond, like lupin and black beans, and different grains as well. It's an ongoing project that keeps surprising and thrilling us at every turn. And it began by listening to stories beyond our own.

Miso is a fermented paste made from a mash of cooked soybeans, koji, and salt. Like vinegar, miso is a two-stage fermentation. First, the fungus *Aspergillus oryzae* is grown on either rice or barley to produce koji (read the koji chapter, page 211) for a deeper understanding of this process). Then the powerful enzymes produced by koji, namely protease and amylase, are harnessed to dismantle the protein and starch in another substrate (traditionally soybeans), cleaving them into amino acids and simple sugars, respectively. Wild yeast, lactic acid bacteria, and acetic acid bacteria also add to the fugue of flavors as the miso ages.

It's difficult to fathom the stroke of creative genius—and luck—that inspired some brave soul to combine moldy rice

271

with cooked soybeans, let it sit for months, and then taste it. We are forever indebted to that act of culinary audacity, as miso is by far the most astounding transformation in the world of fermentation. Time, bacteria, and fungus conspire to convert one-dimensional, everyday staples into arresting new concoctions. One seemingly defies the laws of physics by creating wholly new matter out of thin air. *I didn't put bananas or nuts in here! How can this taste so strongly of bananas and nuts!?*

The full story of how miso came to be is intertwined with the domestication of the sacred soybean, the complicated history between China and Japan, and the dharmic philosophy of nonviolence.

First, the soybean.

As with almost every ancient civilization, China's early existence hinged on the domestication of nutrient-rich crops. What maize was to Mesoamerica, or chickpeas to the Middle East, soybeans were to East Asia. There is no more efficient way to produce protein than to grow soy. It yields nearly twenty times more protein per hectare than grazing cattle or using the land for growing fodder. Of the twenty amino acids that our bodies require to function, there are nine that we can't produce on our own. Soybeans are one of the few plant foods on earth that contain all nine of these essential amino acids.

The earliest evidence points to the cultivation of small, wild soybeans about 7,600 years ago in northern China. The selective breeding of soybeans for increased size began at least 5,000 years ago in China, but may also have started around the same time in Japan. Regardless of its true origins, soy has proven absolutely essential to the region's foodways. In Chinese lore, the "God Farmer" Shennong, a mythical deity, is said to have proclaimed five crops as sacred: rice, wheat, barley, millet, and soybeans.

But nutritive value notwithstanding, it wasn't until soy met fermentation that its true culinary potential was realized.

273

Before miso, there was *jiang*. Jiangs (translating roughly to "pastes") encompass a large array of Chinese condiments and ferments—many that don't contain fermented soybeans at all. In fact, the oldest jiangs were supposedly made from fish or meat, and resembled a sort of thick hybrid between garum (see the chapter on garum, page 361) and miso. As practices of animal husbandry improved through the ages, the imperative to preserve the valuable nutrition of wild meats eased bit by bit, causing the main protein source within jiangs to shift from animals to vegetables over generations. Descendants of these ancient jiangs persist today as well-known condiments, including hoisin, oyster sauce, and fermented black bean paste. Fermented black soybeans, or *douchi*, may be one of the first fermented soy products to come out of China, with references dating as far back as 90 BCE.

China's closest living relative to miso is *huang jiang* ("yellow paste"), where soybeans are steamed and mixed with half their weight in wheat flour before being pressed into bricks and placed on reed mats to undergo wild fermentation in the open air. After a couple of weeks, wild molds that have grown on the surface of the bricks are brushed off and the bricks are mixed with a saline brine, further fermenting into a viscous, salty paste.

Glycine max,
aka the soybean.

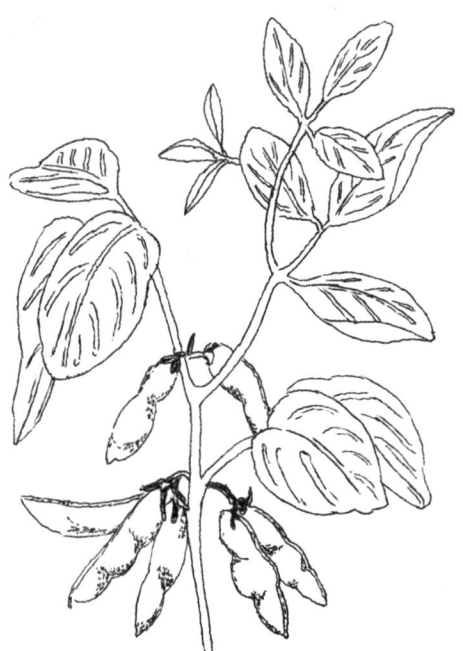

When Chinese Buddhist monks arrived in Japan in the sixth century to "enlighten" the people of the island nation, they brought jiangs with them. The Japanese would absorb the idea of fermenting soybeans and run with it.

In the seventh century, in accordance with the recently introduced dharmic belief in nonviolence, the Japanese emperor Tenmu banned the consumption of farmed animal meat. The mandate, which stayed in effect for more than six decades, created a void in people's diets that had to be filled by vegetarian sources of protein. Soybeans, in the form of fresh edamame, tofu, and miso, took center stage, along with rice and other grains. As miso's importance grew, so too did its utility and variety. For starters, Japanese miso makers brought greater control to the fermentation process by first growing mold on grains, before introducing soybeans to the process.

Miso making became a specialized industry. Early miso (*hishio*) was more a gloopy mash than a thick paste. Over centuries, recipes were refined and fractured into local specialties. In Japan there are dozens upon dozens of varieties of miso, made in myriad styles. There are a number of variables to play with in the miso-making process, each capable of having a profound effect on the outcome: the specific strain of *Aspergillus* mold used to inoculate the koji; the type of rice or barley upon which the koji grows; the method of cooking the soybeans; and the length of time and conditions in which you age the miso. The results can range from red and earthy *aka* miso; to rich, chocolaty, salty *hatcho* miso; to sweet *saikyo* miso. In 2015, when the Noma staff had the chance to spend several months traveling through Japan for a pop-up stint in Tokyo, we felt utterly spoiled for choice.

And all this is to say nothing of the soybean-paste traditions in other parts of Asia. Korea, for instance, has its own broad lineage of *jang*s that developed contemporaneously with misos. Like Chinese jiangs, *jang* is an umbrella term that covers a bevy of fermented products—many but not all made from soy. *Cheonggukjang* is a quick-fermented, chunky analogue to miso, made with the aid of the bacteria *Bacillus subtilis*. *Doenjang*,

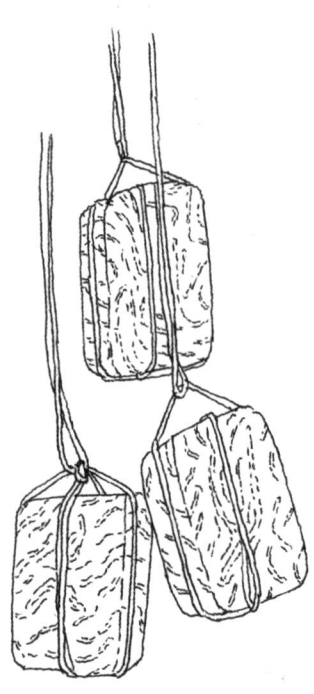

Meju are compacted bricks of fermented soybeans that serve as the basis of several Korean ferments.

Historical Methods and Hand Taste

on the other hand, is a more labor-intensive soy-based ferment that bears striking resemblance to the ancient *huang jangs* of China. It begins with *meju*—dried soybeans that are cooked until tender and then pressed in a wooden box until they firm into a brick. The brick is then removed and wrapped in rice hay, where bacteria and molds endemic to the hay—including wild *Aspergillus*—further ferment the meju brick for two months. Finally, the meju is transferred to earthenware pots, mixed with a saltwater brine, and left to ferment for a year. The resulting liquid is akin to a much funkier soy sauce, called *ganjang*. The fermented solids are doenjang, which is often allowed to age for several more years on its own. *Gochujang*, one of our all-time favorite ferments, also begins with meju, but contains a sizable addition of chiles and glutinous rice flour.

Fermented bean pastes also spread throughout Southeast Asia. In Thailand, we find *tai jiew*, much wetter and funkier than miso. Indonesia has *tauco*, which is quite sweet from the addition of palm sugar. And Vietnam produces the less viscous *tuong*, which you may have consumed as a dip with summer rolls. The fact that this technique has spread throughout Asia—and even as far as chilly Copenhagen—speaks to its tremendous appeal. Miso is contagious.

At Noma, even though we do our best to stay in the vanguard of culinary innovation, we feel a deep responsibility to be respectfully informed by history. We find there's tremendous value in understanding tradition before toying with it. Our team has paid many visits to miso producers in Japan, both large factories and small artisans, gathering invaluable experience that has helped us find our own way. Before we get into the particulars of making a Nordic version of miso, let's take a gander at what miso production looked like for much of its history.

In the early days, almost everything used to make miso was built from wood: spades, trays, large vats, and the buildings themselves were crafted from hardwood, usually Japanese cedar. Massive iron cauldrons would boil the water to steam rice and soybeans held in straw baskets. Cooled rice was spread on a large table and inoculated with a fine dusting of *Aspergillus oryzae* spores.

Workers used spades to turn the rice, ensuring that the spores took hold evenly, then transferred it to cedar trays, which were stacked in a warm, humid room (a *koji muro*).

Soybeans were cooked until tender and crushed beneath men's feet. Workers would then mix the prepared koji and salt with the beans, then ferry buckets of the mixture up stepladders to be emptied into huge cedar fermenting vats. Lids weighted down with heavy rocks would compact the mixture, removing air and leading to more uniform fermentation. Depending on the variety, the miso would age in its cedar container for anywhere from one to three years.

As the miso fermented, salty, umami-rich juices would flow to the top and form a puddle. This liquid came to be known as *tamari*. It's usually less salty and more viscous than its descendant, shoyu. In Chinese, it's called *jiang you*. You probably know it best as soy sauce. The vast wooden warehouses used to store the miso lacked much in the way of temperature control. Fermentation would slow to a crawl in the winter and speed up again in the summer. No two batches were ever identical; each was a product of the specific time and unique conditions under which it fermented.

Now, before you write this off as a quaint but pointless tale of the old days, consider the wisdom of a man named Edward Norton Lorenz. Lorenz served as a weather forecaster in the US Army during the Second World War before earning a doctorate in meteorology from MIT upon his return to the States. His extensive work in the field of weather prediction made him understandably wary of linear statistical methods—that is, the idea that future occurrences could be direct extrapolations of what's happening now. Lorenz knew that the weather operates on very nonlinear phenomena. In a paper published in a meteorological journal in 1963, he wrote: "Two states differing by imperceptible amounts may eventually evolve into two considerably different states . . . Should there be any error in observation, a prediction of the state in the distant future may well be impossible."

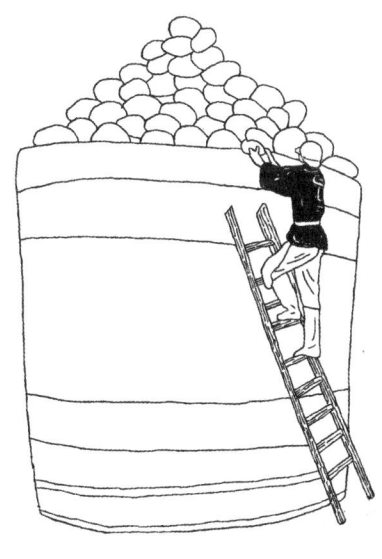

Massive cedar vats known as *kioke* can hold thousands of kilograms of miso at a time.

277

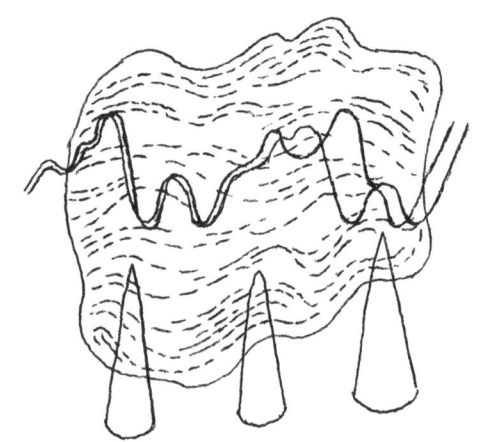

"Hand taste" describes the distinct character of a ferment imparted by its maker and the time and place in which it was made.

Lorenz's thinking would become the foundation of chaos theory. He's often credited with coining the term "butterfly effect," which describes how complex systems with innumerable points of minute differentiation can evolve in dramatically different ways, given enough time. Put another way, something as insignificant as a butterfly flapping its wings can create disturbances that result in a tornado weeks later. The world of fermentation gives us a real-life look at this principle in action. Whether you're aging whiskey or vinegar, brewing sake, or making miso, the more complex the process, the greater the effect small differences at the outset of the process will have. And the longer it takes to ferment something, the more pronounced those differences become.

Korean artisans often speak of "hand taste" (*son-mat*) as an irreplicable quality imbued by individual cooks to their food, a quality absent from factory-made batches. Hand taste, in essence, is chaos theory at work. Minute differences in how miso is made and aged, that day's or hour's population of bacteria on the miso maker's skin and clothes, random variations in temperature, air pressure, or humidity, have such an outsize effect on the ferment's development that they ensure no two batches will ever be exactly the same. It's how cooks and artisans stumble into new flavors and new creations. And it's what makes fermentation unpredictable and thrilling.

Making Our Peaso

Now that we understand the outsize effect that minor alterations can have on ferments, let's take a moment to detail the major steps of how we make our peaso:

1. Inoculate steamed barley with koji spores and allow them to grow for 2 days in a fermentation chamber (see "Building a Fermentation Chamber," page 42).

2. Soak, rinse, and boil yellow peas. Grind or blend the peas with the koji in a 3:2 ratio (by weight).

3. Add salt (4% by weight); then, if necessary, adjust the moisture content of the mixture using a 4% salt brine.

4. Pack the mixture tightly into a fermentation vessel and sprinkle salt over the surface to prevent unwanted mold growth. Weight the mixture down, cover the vessel, and leave it to ferment at a temperature of 22° to 30°C/ 72° to 86°F for at least 3 months.

5. Harvest any tamari that has pooled on top of the miso and scrape away any surface mold. Store the miso itself in airtight containers in the refrigerator.

If you stick to the recipe outlined on page 289, you should be able to produce peaso that's very close to the one we make at Noma. But if you'd like to experiment with the recipe on your own, it's good to have a solid grasp of what's happening in your fermentation vessel and the main factors affecting the final outcome. Having a firm understanding of the following control points will allow you to make adjustments to your own peaso.

Salt Content

Salt content is by far the most important measure in the direct control of the miso/peaso process. As illustrated in the lacto-fermentation chapter (page 55), salt can be used to prevent potentially malevolent microbes from gaining a foothold. But Noma is a decidedly unsalty restaurant, so we try to keep our salt content as low as possible without exposing our peaso to risk.

The more you understand about what goes into making misos, the more you'll be able to fine-tune their outcome. What seems daunting at the start will quickly become a set of tools to use to your advantage, with a little practice.

After many trials, we've settled on 4 percent as the ideal balance between microbial activity and salinity. (Japanese misos, at the very lowest, sit at 6 percent.) We would not suggest going lower than 4 percent salt, as there's no guarantee that unwanted microbes won't propagate over the long aging process.

A higher salt content of 8 to 10 percent will inhibit the growth of yeast, acetic acid bacteria, and, to a lesser extent, lactic acid bacteria (LAB). Now, you may wonder if that leads to a less interesting finished product, given how we've used these microbes to produce flavors in other ferments. But the complexity lost by limiting yeast and bacteria is made up for by long aging times. Over the course of months or years, extremely slow Maillard reactions create alluring and complex flavors while slowly browning the miso. (See "Really Slow Cooking," page 405, for more on the Maillard reaction.) A very salty miso may be less interesting early in its life, but it changes and improves as it ages.

If you ever find yourself producing a batch of peaso that tastes unbalanced, try upping the total salt content by 2 or 3 percent (no other amounts in the recipe will have to change) and aging it for a month or two longer. While you may lose some of the complexity offered by LAB and yeast, the peaso will reach greater deliciousness down the road.

Moisture Content

The wetness of your peaso as it's going into the fermentation vessel is extremely important. If you pack peaso away too dry, the mixture won't be fluid enough to allow for biological or chemical processes to work effectively, in the same way that freezing and dehydrating halts spoilage by holding everything tightly in place. On the other hand, if the mixture is too wet, microbial and enzymatic activity will become far too vigorous. You effectively end up with multiple microbes running amok. A small amount of LAB-produced lactic acid adds a pleasant brightness to peaso, but too much can make it taste completely off-kilter. LAB are only marginally impeded by salt content, and thus need to be corralled by regulating moisture. (More

281

282

Octopus tentacles that have been baked
underground in a crust of masa are
served with Noma's take on *dzikilpak*, the
traditional Mayan salsa, made here with
pumpkin seed miso.

accurately, we're regulating "water activity," or how much
water is unbound and freely available.)

It may take a bit of trial and error to understand the full impact
that water content can have on your peaso, but as a heuristic
guide, if you clench a fistful of miso mixture that's ready to go
into the vessel, it should form a dense ball in your hand. If it
oozes like hummus, it's too wet. If it crumbles, it's too dry.

Humidity

Age your peaso in an environment with an ambient humidity
of 65 to 75 percent. Any lower and the peaso may dry out,
leading to the aforementioned problems with moisture
content, or become too salty as water evaporates into the
air. At the opposite end of the spectrum, peaso held in a very
damp environment can also go awry. For instance, if the
peaso were to come into contact with condensation due to
humidity, it could lower the salinity of the mixture and open
the door for spoilage. It would require a lot of moisture in
the environment for that to happen, but damp basements or
poorly insulated storage areas exposed to persistent rainfall
should be avoided.

Age of Your Koji

It's crucial that the koji you use in your peaso be healthy and
potent, with thick mycelium binding the grains together. The
more fragrant and sweeter the koji, the better your peaso will
be. During koji's short life cycle, the fungus goes through many
different stages. If the koji is too young, it won't have produced
the enzymes needed to break down the proteins in the yellow
peas. A koji that is harvested too early will also be lacking
sweetness that contributes to the overall flavor of the peaso.

On the other hand, fungus that has been grown too long and
gone to spore will have a considerably different flavor from
properly harvested koji, in the same way that garden greens
change dramatically once they've gone to seed. Sporulated
koji uses its enzymes to consume sugars in order to fuel the

283

production of its spores. Less sugar in the koji means less sugar in the miso to take part in the Maillard reaction responsible for the development of complex flavors as it ages.

In our experience, 44 to 48 hours from the time of initial inoculation is the window in which you want to harvest koji for peaso. Finally, if your koji shows any signs of being infected by unwanted mold or other microbes, don't use it for peaso. Remember, your fermented products are only as good as the ingredients that go into them.

Temperature and Time

The warmer your peaso, the faster it will ferment. Peaso held in environments cooler than room temperature, like a cellar, will ferment noticeably more slowly. Maillard reactions will also accelerate in warmer environs, creating more toasted flavors in your final miso.

The enzymes produced by koji are quite efficient when catalyzing biochemical reactions when held at 60°C/140°F. Chances are, you won't be fermenting your peaso in an insulated room heated to a constant temperature of 60°C/140°F, nor would you want to—it would end up tasting burnt very quickly. At 28° to 30°C/82° to 86°F, you'll be emulating Japanese summers, and can reap the benefits of faster fermentation and more prolific Maillard reactions.

However, even at moderate temperatures, over a very long time, peaso can begin to deteriorate or even burn. Once it ferments for, say, a year, it's likely not going to keep getting better. Trying to guess when something is at its peak is difficult. But with experience you'll begin to see the signs. If the peaso begins to grow bitter, dries out, or darkens significantly, pay heed. It's probably not going to improve with age at this point.

Pressure and Exposure to Air

Placing weights on top of peaso forces out air pockets that would otherwise be ideal breeding grounds for acetic acid bacteria or mold to create unwanted flavors. The weight should be at least equal to half the weight of the peaso in the vessel. However, even when pressed beneath a weight, peaso isn't sealed off completely from air—that's by design. The microbes at work produce gases that need to vent. If the peaso were sealed airtight, the gases would be reabsorbed and taint the flavor. A breathable cloth laid over the top will allow you to vent these gases, while keeping out larger agents of spoilage like flies and their maggots.

If your peaso tastes somewhat acrid, acidic, or alcoholic, it may be that it was packed too tight or not allowed to vent properly. This can be partially remedied by cooking the peaso in a pan for five minutes, stirring often with a spatula. The unwanted aroma and flavor molecules are fairly volatile, meaning they'll dissipate once heated. Unfortunately, there's no way to get rid of these flavors completely. Live and learn.

Additional Flavorings

Miso is an ideal receptacle for scraps. Adding the pulp of a juiced vegetable, seed husks, or leftovers can lead to remarkable flavors a year down the line. For example, each summer, we harvest wild beach roses and blend them with a neutral oil to infuse it with their flavor and aroma. The pulp that remains after we press the oil still has tons of value. We add some of it to our basic peaso recipe—5 percent of the total weight, plus additional salt—and the transformation is incredible. Crack open a bucket and robust floral scents fill the room. You create something profoundly delicious by diverting a waste stream. Don't be afraid to experiment. Vegetable tops, pine needles, the strained solids from a batch of garum, fruit peels—they can all lead to delectable discoveries.

That said, we don't advise any ingredient additions that exceed 20 percent of the total weight of the peaso. The fermenting legumes themselves are what make miso amazing. If there aren't enough proteins available during fermentation, your miso additions will simply become salty and fester. They may not go bad, but we're striving for amazing here.

285

Yellow peaso is Japanese in spirit, Nordic in execution,
and indispensable in our kitchen.

Yellow Peaso

Makes 2.5 kilograms

800 grams dried yellow split peas
1 kilogram Pearl Barley Koji (page 231)
100 grams non-iodized salt,
    plus extra for sprinkling

Miso made with yellow peas was a revelation at Noma. When the first batches of "peaso" made their way around the kitchen, many of the cooks working at the time became completely enamored of the new concoction, and found any excuse to use it wherever they could. It's boundlessly versatile, and once you realize the potential of a spoonful here or there, you'll never be able to go back to a time when you cooked without it.

Making peaso is ostensibly straightforward, but because it ferments for such a long time, you'll need to pay attention to the process, because small details will lead to big differences in the outcome. This recipe contains detailed instructions, but please be sure to read about the various points of control explained in the preceding section as well.

Equipment Notes

First, you'll need a means of grinding the peas into a coarse meal, such as a meat grinder or food processor.

The peaso will ferment in a nonreactive (glass, plastic, ceramic, or untreated wood) fermentation vessel of about 5-liter capacity. You'll need to press the peaso down with weights and cover it with cheesecloth or a clean kitchen towel. We also recommend you wear sterile gloves when working with your hands and that all your equipment be thoroughly cleaned and sanitized (see page 36).

The peas should be *just* cooked—easily squished between two fingers but not mushy.

Break up the koji in a food processor.

In-Depth Instructions

Soak the dried peas in cold water for 4 hours at room temperature to rehydrate them. Use double the peas' volume in water as they'll absorb a good amount of it; you don't want any peas left dry above the water line. Once the peas have soaked, drain them, place them in a large pot, and cover again with double their volume of cold water. Bring the water to a boil, then reduce the heat to a bare simmer, skimming away any starchy foam that rises to the surface. Cook for 45 to 60 minutes, stirring every 10 minutes or so, until the peas are soft enough to crush between your thumb and forefinger without applying much pressure.

Drain the peas and spread them on a baking sheet to cool to room temperature. Once the peas are cooled, weigh them. You should have close to 1.5 kilograms, but the amount of water the peas absorb in the soaking and cooking process will always vary. If you've got more than 1.5 kilos, you can set the extra peas aside for another use. If you have less than 1.5 kilos (or if you want to use any excess), you'll simply need to adjust the ratio of the other ingredients. The amount of koji needed is 66.6% of the weight of the cooked peas; the salt is 6.6%. If, for instance, you wind up with 1.3 kilograms of cooked peas, reduce the amount of koji from 1 kilogram to 866 grams, and the salt from 100 grams to 86 grams. These are exact ratios that you should adhere to if you want your peaso to turn out the way we've intended.

To blend or mash the peas with the koji, the best option is to use a clean, sanitized meat grinder. First, put on a pair of latex or nitrile gloves, place the cooked peas into the hopper, and grind them with a medium die into a very large bowl or container. Next, grind the koji and add it to the peas. (Alternatively, you can use a food processor, but be careful that you don't overprocess things. You're not aiming for a puree—a coarse meal will suffice. As a last resort, if you don't have a meat grinder or food processor, you can mash the peas in a large mortar and pestle and crumble the koji by hand.)

Give the ground peas and koji a good mixing by hand, then check the texture and moisture content: Squeeze a small handful of the mixture. If it easily forms a compact ball, you're good. If the mixture crumbles, it's too dry and you'll need to add some water to hydrate it. However, it's vital that you maintain the 4% salt ratio, so any liquid you add to the mixture should have the same salt content. Make a quick 4% salt brine by blending 4 grams salt into 100 grams water with a handheld blender or whisk until the salt has completely dissolved. Add a little bit at a time to the pea mixture, until you've achieved the proper texture.

If the mixture oozes out of your hand when squeezed, it's too wet; the peas may have been overcooked or improperly drained. Too wet is more difficult to correct than too dry, but not impossible. Spread the mixture on a parchment-lined baking sheet in a thin, even layer and dry in an oven or a dehydrator at a low temperature (40°C/104°F), giving it the squeeze test frequently, until it reaches the texture you're looking for.

Once you're happy with the texture of the mixture, add the salt and mix it thoroughly once more. (Remember to adjust the quantity of salt if your peas weighed anything other than 1.5 kilograms.) Now it's time to pack away the peaso.

At Noma we ferment our peaso in food-safe plastic buckets, although a glass or ceramic jar will work well, too. If you have access to a cedar fermenting vat, feel free to use it, but be sure that the wood is untreated.

Working one gloved handful at a time, transfer the peaso to the fermentation vessel and pack it as tightly as possible. Start at the edges of the container, forcing any air out, then work your way toward the center. Punch the mixture down with your fists after each addition to ensure that it's well packed. Smooth and flatten the top of the peaso and lightly sprinkle the surface with salt to help prevent mold from forming. Place a sheet of plastic wrap over the top, in direct contact with the peaso, making sure it reaches all the way to the edges. Finally, wipe down the walls of the container with a clean paper towel.

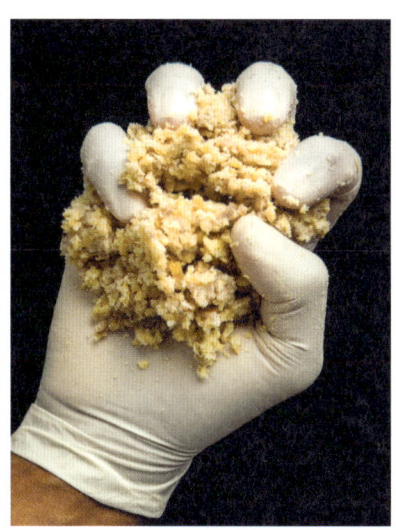

Peaso mixture that is too dry crumbles when squeezed.

Peaso mixture that is too wet squishes and oozes when squeezed.

291

Yellow Peaso, day 1

Day 14

Day 30

Now you'll need to weight down the peaso. As the peaso ferments and yields tamari, the weight will keep the mixture submerged in liquid in the same way that a lacto-ferment like sauerkraut rests beneath its juices. If you like, you can purchase specially designed fermentation weights online that will fit the circumference of your fermentation vessel. Otherwise, the simplest method is to use a flat dinner plate that fits inside your fermentation vessel snugly. If you're using a plate, bear in mind that the plate will sink over time and that eventually you'll need to remove it, so be sure it doesn't fit *too* snugly or you won't be able to take it out. Place the plate right-side up on top of the peaso and press it down with your hand. Now procure a rock, a brick, or a few cans that weigh roughly half as much as the peaso—about 1.5 kilograms. Place the weights in plastic bags to keep things sanitary and distribute them evenly over the plate.

You can also forgo the plate method in favor of zip-top or vacuum bags filled with a total of 3 liters of water. (You need to use more weight in this method because some of the pressure will be spread outward against the walls of the container and not directly down on the peaso.) Double-bag the water to avoid leakage and lay the bags over the peaso.

Cover the container with a clean kitchen towel or cheesecloth and secure it with a couple of large rubber bands.

Your peaso will do fine fermenting on your kitchen counter at room temperature, but at Noma we age our peaso in a dedicated room held at 28°C/82°F for about 3 months. It should age well in either scenario, although an extra month might be necessary at room temperature, and you can certainly leave it even longer than that, if you like. We've run really long peaso experiments at Noma, and the results have been interesting. It gets much richer, with darker, earthier tones, the longer it ferments, but we really prefer the versatility of 3-month-old peaso.

Day 60

Day 90

Check the peaso's progress after 3 or 4 days. It won't look all that different from when you started. If anything, it will be slightly more aromatic. If that's the case, it's going well. If you notice it souring like a lactic ferment, with a lot of tamari pooling on top, it means your mixture was too wet and you'll have to start over. If you're fermenting in a clear container, you may notice small air pockets forming throughout the peaso. These are part of the normal fermentation process; they will subside over time.

After the first couple of weeks, open the peaso every week or two to check its progress. Be sure to wear gloves to avoid introducing contaminants. At some point, you may find white mold growing on the surface. That's totally fine. In our experience, it's usually a patch of koji that has managed to grab a foothold on an exposed portion of the mixture. But even if it's another mold, if the miso is packed tightly, the mold won't be able to penetrate the surface. When you need to taste the peaso, scrape a bit of the mold to the side to get underneath, but don't remove it completely until you harvest the whole batch, lest more come back in its place.

The peaso is finished when the texture has softened significantly, the taste of salt has subsided slightly, and all manner of sweet, nutty tones have emerged—usually somewhere between 3 and 4 months. It should have a mild acidity without being overly sour. The texture of the peaso will be slightly nubby, so if you'd prefer a very smooth paste, blend the peaso in a food processor—add a bit of water to help it spin, if necessary—after which you can pass it through a tamis if you're looking for a truly velvety texture.

You can pack the peaso into airtight jars or containers and store them in the fridge for use within the month. Any longer than that and we suggest you store it in the freezer to keep its flavor freshest; just pull it out as you need it.

293

1. Soak, rinse, and cook peas in at least double their volume in water.

2. Drain and cool the peas, then process or grind them into a coarse meal.

3. Measure out 66.6% of the cooked peas' weight in koji, and process or grind it as well.

4. Thoroughly mix together the peas and koji with 4% of their combined weight in salt.

5. Pack the mixture tightly into a fermentation vessel.

6. Sprinkle the top of the peaso with salt to help prevent surface mold growth.

7. Cover the surface with a layer of plastic wrap.

8. Weight down the peaso.

9. Cover the vessel with a breathable cloth and secure with a rubber band.

10. Allow the mixture to ferment for a minimum of 3 months.

11. Scrape away any mold that may have formed on top of the peaso.

12. Remove the peaso from the vessel and pack it into clean containers and store it in the refrigerator or freezer.

Ice-clarifying a blend of peaso and water produces a savory liquid that will later reduce into one of the most potent and valuable ingredients in the Noma kitchen.

Suggested Uses

Peaso Tamari Reduction

One of the first things to grab our attention when we began making peaso was the incredible tamari that would pool on top during the fermenting process. It holds a perfect syrupy balance of umami, saltiness, sweetness, and acidity. The problem was, there was never enough of it, so we devised this workaround.

In a blender, thoroughly pulse 130 grams peaso into 860 grams water and transfer the mixture directly to a freezer-safe plastic container. Cap it with a lid and freeze solid overnight.

The following day, place the brick in a cheesecloth-lined sieve and set it over a bowl to catch the liquid as it thaws. When the mass has no more liquid to surrender, discard the pulp and transfer the liquid to a small pot on the stove. Reduce the liquid slowly until it can coat the back of a spoon. Once cooled, peaso tamari reduction will keep extremely well, covered, in the fridge.

You can combine the peaso tamari reduction with finely minced herbs, like cilantro or parsley, to make an out-of-this-world herb paste that is great on steamed vegetables. Or you can whisk it together with an equal quantity of clarified butter for a marinade that was born for basting pork belly. The absolute favorite meal in the Noma test kitchen is a bowl of freshly steamed rice, drizzled with peaso tamari and topped with a spoonful of trout roe or, if you're feeling luxurious, a few fat tongues of sea urchin.

Roasted-Garlic Oil's Ideal Companion

The ample sugars in miso caramelize beautifully on the grill, and combining it with a flavorful oil will provide the fat necessary to let those sugars bubble and brown without drying out. Roasted-garlic oil is a perfect partner for peaso. To make roasted-garlic oil, peel and crush the cloves of 1 head of garlic, then place them in a small pot. Cover with twice their volume in neutral vegetable oil. Place the pot over low heat and watch carefully until the cloves start bubbling rapidly. Reduce the heat as low as it will go and cook for 1 hour, then remove the pot from the heat and let the oil cool down to room temperature. Once cool, transfer the cloves and oil to a covered container and refrigerate overnight. Strain the oil (saving the garlic cloves for whatever your heart desires). Store the oil and garlic cloves in separate airtight containers in the fridge. The oil will last for weeks; the cloves for a few days.

Garlic-Peaso Cabbage

To see the powerful combination of garlic oil and peaso in action, take it for a spin with some cabbage leaves. Briefly blanch the leaves in a pot of flavored stock (like Koji Stock, page 254), then shock them in ice water. Pat the leaves dry and smear one side with a thin layer of blender-smoothed peaso (see the recipe for Peaso Butter, page 300, for instructions on how to do this). Drizzle roasted-garlic oil lightly over both sides and grill the leaves peaso-side down over very high heat until the peaso caramelizes and the edges begin to crisp. Serve the grilled leaves all on their own, or roughly chop them for a salad with thick sourdough croutons, Sun Gold tomatoes, and anchovies.

299

Garlic-Peaso Grilled Beef

Garlic oil and peaso also play very nicely with beef. Make a marinade by whisking together 1 part roasted-garlic oil with 3 parts blender-smoothed peaso (see Peaso Butter, below), by volume, until the mixture emulsifies. Slather a thick slab of your favorite cut of beef with the mixture and leave it to marinate in the fridge for a few hours before grilling. (Thick cuts are best, because the salt in the peaso will actually begin to cure the beef.) Don't even bother to wipe off the marinade, as it's a powerhouse of flavor and will help form a tasty crust.

Peaso Butter

Peaso butter is one of the easier and most useful applications for peaso. To make it, you'll first want to smooth out your peaso. Place 100 grams peaso in a blender and spin it until smooth. Depending on your blender, you may need to add a bit of water to help it catch—but not too much or you'll dilute the flavor.

Once you've got smooth peaso, pass it through a tamis. This isn't absolutely necessary, per se, but as Thomas Keller once put it, the tamis creates "the texture of luxury." Next, whisk the peaso into 400 grams room-temperature butter until fully incorporated. From there, you can roll the compound butter into a tight cylinder on a sheet of plastic wrap and store it in the fridge, so you can slice off a puck whenever you need to baste chicken in a cast-iron pan, or melt some over whipped potato puree.

Slicing pucks of peaso butter.

301

Rose Peaso

Makes 2.5 kilograms

1.5 kilograms cooked split yellow peas
  (see page 290)
950 grams Pearl Barley Koji (page 231)
125 grams wild rose petals
100 grams non-iodized salt, plus extra
  for sprinkling

At Noma, making miso is often a way for us to funnel what would otherwise be waste into a new product. This recipe was designed to use up leftover pulp we had from blending wild rose petals to make rose oil. We don't expect you to work in the opposite direction, making a completely separate product in order to get the by-product, so we've adapted the recipe to use fresh ingredients, which function just as well.

The in-depth instructions for Yellow Peaso (page 289) serve as a template for all the miso recipes in this chapter. We recommend you read that recipe before starting in on this one.

Using a clean, sanitized meat grinder or food processor, grind the cooked peas and koji into a coarse meal. Gather the rose petals, stack and roll them into a loose cylinder, then slice them into a chiffonade as you would basil leaves. Fold the roses in with the peas and koji, and add the salt. Wearing gloves, mix everything thoroughly.

If necessary, adjust the texture and moisture content of the rose peaso mixture by adding a little bit of 4% salt brine at a time, until you can squeeze the mixture into a tight ball in your fist. (See the yellow peaso recipe for more detailed instructions.)

Rose Peaso, day 1

Day 30

Day 90

Pack the rose peaso mixture tightly into the fermentation vessel. Smooth and level the top, wipe the insides of the vessel clean, and sprinkle the surface with salt. Weight down and cover the peaso, according to the directions outlined in the yellow peaso recipe. Let the peaso ferment at room temperature for 3 to 4 months. Once the peaso has finished fermenting, you can blend it with a bit of water until smooth and then pass it through a tamis for a finer texture. Pack the finished peaso in airtight jars or containers and store in the refrigerator or freezer.

Other Floral Misos

The basic concept of this recipe—yellow peaso with the addition of 5% by weight of a floral ingredient—can be adapted to create a whole slew of amazing flavored peasos that end up being something far beyond the simple sum of their parts. The fermentation process turns highly aromatic flavors into something completely new. Without going into the details of every single one, we offer you a small list of our "greatest hits."

- Cacao
- Cherry blossom
- Douglas fir
- Elderflower
- Hibiscus
- Lemon verbena
- Marigold
- Meadowsweet
- Orange blossom
- Pineapple weed
- Pink pepper leaves
- Plum kernel
- Rose geranium
- Vanilla pod
- Yuzu rind

Caramelizing rose peaso in butter.

Suggested Uses

Dressing for Summer Fruits

The intense floral bouquet of rose peaso plays remarkably well with dark fruits like plums, blackberries, and mulberries. Try blending a few big spoonfuls of rose peaso and passing it through a tamis. Thin the peaso out with just enough water to achieve the consistency of a loose yogurt. Toss berries or bite-size pieces of stone fruit in just enough of the rose peaso sauce to coat them. Finish with the lightest sprinkling of flaky sea salt.

Mashed Root Vegetables

For an ideal accompaniment to white fish like turbot or halibut, cook a batch of Jerusalem artichokes or new potatoes in a pot of salted water. Drain, return the vegetables to the pot, and smash them lightly with a fork. Throw in a big knob of butter along with the same amount of rose peaso. Stir until everything becomes creamy, and season with salt to taste.

Caramelized Rose Butter

Rose peaso begins as an earthy product with delicate over-tones, but once it's caramelized it finds even more potency as the base for a warm vinaigrette. Put 40 grams rose peaso and 200 grams butter in a small pot over medium heat. As the butter melts and begins to clarify, stir constantly with a silicone spatula or small whisk—the rose peaso will sit at the bottom and burn if you don't keep it moving. After about 20 minutes, the butter should be clarified and the rose peaso will have sweetened and browned. Remove the pot from the stove and spoon the rose butter over just-cooked crab, lobster, or shrimp.

Rose peaso is a perfect foil for summer berries.

Ryeso marries Nordic ingredients with Japanese techniques.

306

Ryeso

Makes about 3 kilograms

1.8 kilograms Danish-style rye bread
1.2 kilograms Pearl Barley Koji (page 231)
120 grams non-iodized salt, plus extra
for sprinkling

In trying to come up with a local analogue for the mighty soybean from which we could make a uniquely Nordic miso, our minds went in all kinds of outlandish directions. For instance, if koji can break down the starches in whole grains, then why not the starches in something made from grains? Like bread? We tested the idea with that most Danish of baked goods, rye bread, and to our delight, it worked spectacularly. The resulting miso—or "ryeso," as we dubbed it—was an even sharper representation of Noma's Danish home than our yellow peaso, with all the warm and satisfying notes of freshly baked rye bread stretched out across a salty plane of deep umami and tanginess.

The in-depth instructions for Yellow Peaso (page 289) serve as a template for all the miso recipes in this chapter. We recommend you read that recipe before starting in on this one.

You can purchase pre-sliced Danish rye bread (much denser and sourer than Jewish rye) in a small block-type package from many grocery or health food stores. And of course, if you have access to a bakery that makes fresh Danish-style rye, by all means use that.

Ryeso, day 1

Day 30

Day 90

Wearing gloves, cut the rye bread into pieces that will be easy for your food processor to manage. Pulse the bread until it crumbles into a coarse meal and transfer to a large sterilized bowl. Next, blend the koji in the food processor.

Add the koji and the salt to the rye crumbs and thoroughly mix the ingredients together by hand. Unlike peaso, where the texture is often spot-on from the start, dryness can be a problem when making miso from bread, and you will almost certainly need to add moisture. Make a quick 4% salt brine by blending 4 grams salt into 100 grams water with a handheld blender or whisk, until the salt has completely dissolved. Add the brine a little bit at a time, until you can squeeze the mixture into a ball in your fist.

Pack the ryeso mixture tightly into the fermentation vessel. Smooth and level the top, wipe the insides of the vessel clean, and sprinkle the surface with salt. Weight the ryeso down according to the directions outlined in the Yellow Peaso recipe (page 289) and cover. Let the ryeso ferment at room temperature for 3 to 4 months. Pack the finished ryeso in airtight jars or containers and store in the refrigerator for a month or freezer for a few months.

Suggested Uses

Ryeso Cream

For a sweet take on ryeso, blend 70 grams ryeso with a hand-held blender until smooth (using a bit of water if you need to help it along), then pass it through a tamis, resulting in a smooth, uniform paste. Whisk together with 70 grams simple syrup (equal parts muscovado sugar and water, by weight, that have been boiled and cooled), then add 250 grams double (or heavy) cream and a pinch of licorice powder. You don't need to whip it, just whisk to combine. Ryeso cream can be paired with cakes and confections of all sorts, but at the restaurant we love it on fresh berries at the height of the season, dressed in their own fermented juices (see Lacto Blueberries, page 97). If you're able, finish it all off with a few marigold flower petals and leaves.

Blended with sugar, cream, and a pinch of licorice powder, ryeso can be turned into a rich, sweet sauce.

309

Ryeso Tamari and Ryeso-Mushroom Glaze

Like peaso, ryeso can yield a delicious tamari reduction, with complex, sweet notes of malt inherited from the rye bread. Follow the instructions for Peaso Tamari Reduction (page 298), substituting ryeso for peaso.

Or take things one step further: Ryeso tamari consorts particularly well with dried mushrooms, and the two can become a hyper-flavorful umami bomb that's been a miracle worker at Noma for years. Once you've reached the point in the tamari process where you would ordinarily reduce the mixture, stop to infuse the tamari with mushrooms. For every 500 grams ice-clarified ryeso stock, add 10 grams each of dried cep, morel, and black trumpet mushrooms, along with 25 grams dried kelp. Bring the mixture to a boil, then reduce the heat to a bare simmer and cover the pot tightly with a lid. Leave it on low heat for 2 hours. Strain out the mushrooms and seaweed, pressing to extract as much liquid as possible, then return the ryeso tamari to the stove. Reduce the tamari over low heat until it can coat the back of a spoon.

The resulting glaze is phenomenal painted onto the skin of ducks, pheasants, or quail as they cook over coals. For vegetarians, grill a whole cluster of hen of the woods mushrooms with a bit of melted butter, brushing periodically with ryeso-mushroom glaze. The spongy fungi will soak up the flavor and become a crispy-smoky-juicy-meaty wonder, capable of satiating any carnivore.

Clarified ryeso stock simmered with mushrooms and kombu leads to an incredibly savory glaze for meat or vegetables.

311

Maizo

Makes about 3 kilograms

2 kilograms Masa (page 315)
1.3 kilograms jasmine rice koji (opposite)
130 grams non-iodized salt, plus extra
   for sprinkling

There are hundreds of varietals of dried corn waiting to be nixtamalized and fermented.

We developed this ferment while spending several months in Tulum, Mexico, building and executing a Noma pop-up in the jungles of the Yucatán. The idea was to produce a miso by swapping ingredients that played a similar role in disparate cultures. Soybeans are a staple in Japan, just as peas are in northern Europe and corn is in Mexico. But corn isn't simply eaten as corn in Mexico. By boiling the kernels in a calcium hydroxide solution, then grinding them, you get masa: the basis for tortillas, tamales, *huaraches*, *sopes*, and countless other vital Mexican dishes. The soaking process is called nixtamalization, and it effectively breaks down the cellulose in the corn's cell walls, rendering the corn more digestible while simultaneously unlocking nutrients and flavor compounds. And so, we started with masa as the basis for our new ferment, a striking and unexpected coming-together of techniques and traditions that we now lovingly call "maizo."

The intense sweetness of maizo means it caramelizes beautifully on the grill, slathered over ears of corn or pork chops or peaches. We're exceedingly proud of this ferment and how it came about. We think if you take the time to make it, you'll see why.

The in-depth instructions for Yellow Peaso (page 289) serve as a template for all the miso recipes in this chapter. We recommend you read that recipe before starting in on this one.

Maizo, day 1

Day 30

Day 75

If you live somewhere you can buy fresh masa, feel free to use it in lieu of making your own, but don't be tempted to substitute Maseca, a brand of dried, instant masa. The flavor is just not the same. As for the jasmine rice koji, if you've mastered barley koji, making rice koji will be straightforward. Follow the procedure described for Pearl Barley Koji (page 231), substituting an equal weight of rice for barley. Obviously, if you've already made barley koji and have some around, it'll work just fine. We prefer rice koji, but fermentation can be flexible and so can we.

Wearing gloves, break up the masa into a large bowl. Put the rice koji in a food processor or meat grinder to break it up as well. Combine the masa, koji, and salt and mix until thoroughly combined.

Unlike other misos in this chapter, you won't need to adjust the texture or moisture content of the maizo mixture with salt brine. The masa holds a lot more water in its starch than peas or soybeans. As the amylase in the koji breaks up the long chains of starch, all that water is liberated. Thus, maizo always ends up far wetter than when it starts. That said, don't worry about the flavor going awry. The high pH of the masa helps deter LAB and yeast, leaving this ferment to work primarily by enzymes alone.

Pack the maizo mix tightly into the fermentation vessel. Smooth and level the top, wipe the insides of the vessel clean, and sprinkle the surface with salt. Weight down and cover the maizo, according to the directions outlined in the Yellow Peaso recipe (page 289). Let the maizo ferment at room temperature for a little less time than you would other misos—2 to 2½ months. The fruitiness of the maizo doesn't benefit from the earthy tones associated with longer aging times. Pack the finished maizo in airtight jars or containers and store in the refrigerator or freezer.

313

Suggested Use

Fish Cure

The strong floral and fruity tones of maizo make it an impeccable cure for flat white fish, like fluke or flounder. Smear both sides of a couple of thin fillets with a good coat of maizo—the fish doesn't need to be buried, but the entire surface should be covered. Place the fillets in a baking pan and marinate in the fridge for 1 hour. While it's curing, take the time to prepare some accompaniments: julienned scallions, passion fruit seeds, thin slivers of jalapeño, and coarsely chopped cilantro. Remove the fillets from the fridge and scrape off the cure with a spoon, then wipe off anything left with a damp paper towel. Slice the fish on the bias into thin ribbons. Place the slices on a flat plate and top with the prepared accompaniments, a healthy drizzle of olive oil, some sea salt, and the zest and juice of a lime.

Marinate a flounder fillet in maizo for an hour before slicing it thin and serving with herbs and a squeeze of lime.

Masa

Makes about 3 kilograms

1 kilogram dried corn
5 grams calcium hydroxide

Calcium hydroxide is available online and in Mexican markets, sometimes under the names "cal" or "pickling lime." Be sure to use calcium *hydroxide* and not calcium oxide, which is not edible and can be dangerous.

Place the corn and the cal along with 5 liters water in a large pot and bring to a boil, stirring occasionally. Reduce the heat to a lively simmer and cook gently until the corn is al dente and you can break the kernels apart with your fingernail, about 50 minutes. Remove the pot from the heat, cover it with cheese-cloth, and allow it to sit out overnight (or for at least 12 hours). The following day, drain the corn and rinse it gently under cold water for about a minute. Pulse the washed corn in a food processor until it has the texture of a fine meal. Reserve the masa in the refrigerator in an airtight container until needed.

Suggested Use

Tostadas

If you live in a part of the world where you can buy fresh tortillas (or tostadas), more power to you. If not, now you can make your own. Use the palms of your hands to press 30-gram balls of masa between two sheets of plastic until you have disks about 2 millimeters / $\frac{1}{16}$ inch thick. Griddle the tortillas on both sides in a hot, dry pan until they puff up. Transfer the cooked tortillas to a baking sheet and bake them at 140°C/ 285°F until they're completely dry and crisp, about 20 minutes. Spread a generous spoonful of maizo onto each tostada and top with anything your heart desires: slices of avocado, grilled octopus with salsa verde, spiced crickets, or adobo-marinated chicken, to name a few.

315

Hazelnut miso resulted from a need to use up
defatted hazelnut pulp.

316

Hazelnut Miso

Makes about 3 kilograms

1.9 kilograms defatted hazelnut meal
(see Sources, page 448)

1.2 kilograms Pearl Barley Koji (page 231)

120 grams non-iodized salt

If You Can't Cut the Fat, Cut the Time

If sourcing defatted hazelnut meal proves difficult, you *can* successfully make this miso with full-fat hazelnut meal, but it comes with a trade-off. The miso has to be fermented for a much shorter time than it normally would to mediate the risk of an excess buildup of decomposed fats. As with Pumpkin Seed Miso (page 325), an aging time of 3 to 4 weeks will be plenty to begin developing interesting fermented flavors while keeping the taste of rancidity at bay.

Nuts seem an obvious candidate for fermentation as misos. They're high in protein, starchy, and abundant in northern Europe. But they come with a caveat: fat. The first many times we tried hazelnut miso, rancid tones set in before complex fermented flavors developed. The rancidity stemmed from the lipids in the hazelnut breaking down as a part of normal fermentation. *A. oryzae* produces lipase—albeit at far lower concentrations than its other two enzymatic workhorses, amylase and protease—which cleaves fats apart into their constituent molecules (fatty acids).

When fat is whole and fresh, its flavor is delicious and satisfying, hence our intense cravings for and ability to gorge on it. Fatty acids, on the other hand, can strike us as disgusting, because we associate them with decaying (i.e., rancid) fat.

The solution? Scratch the fat. Shortly after we began trying to make hazelnut miso at Noma, the test kitchen acquired a new toy: a nut press. Nut presses grind nuts and separate the pulp from the oil by driving the mash through a heated auger. The team in the test kitchen was after nut oils for the menu, but the fermentation lab saw an opportunity: fat-free nut pulp. It offered the perfect opportunity to build a nut miso without any off-putting fatty acids, and it worked splendidly. Fortunately, you don't need to purchase a large piece of industrial machinery to make this miso. You can find low-fat or defatted hazelnut meal online.

317

Hazelnut Miso, day 1

Day 30

Day 90

The in-depth instructions for Yellow Peaso (page 289) serve as a template for all the miso recipes in this chapter. We recommend you read that recipe before starting in on this one.

Heat the oven to 160°C/320°F. Spread the hazelnut meal onto baking sheets and toast in the oven until lightly browned and aromatic, 20 to 25 minutes. Stir every 5 minutes to ensure that the meal browns evenly. Cool to room temperature on the counter. Reweigh the meal. You want to end up with 1.8 kilograms, but the nut meal will lose moisture, and thus weight, during toasting, which is why we start with 1.9 kilos.

While the hazelnut meal is toasting, grind the barley koji in a food processor until it's well broken up.

Combine the toasted hazelnut meal, koji, and salt in a bowl. Wearing gloves, mix everything thoroughly. Unlike peaso, where the texture is often close to spot-on from the start, dryness can be a problem with hazelnut miso. You will almost certainly need to add moisture. Make a quick 4% salt brine by blending 4 grams salt into 100 grams water with a handheld blender or whisk until the salt has completely dissolved. Add a little bit of the brine at a time, until you can squeeze the mixture into a ball in your fist.

Pack the mixture tightly into the fermentation vessel. Smooth and level the top, wipe the insides of the vessel clean, and sprinkle the surface with salt. Weight down and cover the hazelnut miso, according to the directions outlined in the yellow peaso recipe. Let the hazelnut miso ferment at room temperature for 3 to 4 months. Once the hazelnut miso is fermented, you can blend it with a bit of water until smooth and then pass it through a tamis. Pack the finished miso in airtight jars or containers and store in the refrigerator for a month or freezer for a few months.

Suggested Uses

Onion Salad

Peel and halve golf ball-size sweet onions from stem to root, then toss them lightly in a bit of oil before grilling them facedown over hot coals. Once the faces of the onions caramelize and blacken, remove them from the grill grates and wrap them in foil. Set the foil packet off to the side of the grill to allow the onions to continue to cook until tender but still offer a little bite, about 10 minutes. Remove the onions from the packet and shuck the petals into a bowl. Toss them with a big spoonful of the hazelnut miso, pureed and passed through a sieve, a bit more oil, salt, pepper, and picked thyme and oregano leaves. It's a great side dish as is, but you could also toss the onion petals with a mixture of watercress, dandelion, and arugula.

S'mores

Once you taste it for the first time, hazelnut miso will almost certainly replace any nut butter as your new favorite. And if you think of it as a nut butter, it's easy to come up with ways to use it. For a quick example, smear a spoonful of hazelnut miso onto graham crackers the next time you're making s'mores with (or without) the kids.

Breadso began as a project to use up extra bread,
but it quickly proved to be an extraordinary product in
its own right.

Breadso

Makes about 2.5 kilograms

3 kilograms crustless sourdough bread
Aspergillus oryzae koji tane
(see Sources, page 448)
100 grams non-iodized salt, plus
extra for sprinkling

As with Ryeso (page 307), in this miso we'll be using koji to break down bread. But unlike ryeso, we're going to grow the koji directly on the bread—no rice, no barley. What you're looking for is day-old sliced bread. We remove the crusts for the same reasons we remove the husks from grains—the koji's hyphae can have a hard time getting through. With this recipe, and through the transformative power of mold, you can turn leftovers into deliciousness.

Wearing gloves, use a serrated knife to cut the bread into roughly 2-centimeter / ¾-inch cubes. Run the cubes of bread through a food processor until they crumble and ball up around its sides. Steam the crumbled bread for 5 minutes to moisten it a bit. Remove the bread from the steamer and let it rest on the counter for 10 minutes to cool, while allowing the water time to seep into the bread and hydrate it evenly.

Follow the process described in the Pearl Barley Koji recipe (page 231): Spread out the bread, inoculate with the koji spores, and incubate. The koji should cover the blended bread quite thoroughly within 48 hours. Once the bread koji is ready, weigh out 3 kilograms and pulse it into a paste in a food processor. Transfer to a bowl, add the salt, and mix it thoroughly with gloved hands.

Breadso, day 1

Day 30

Day 90

Adjusting the moisture content and texture of this miso is somewhat more difficult than other misos. The bread koji acts like a sponge, and you won't be able to judge how saturated it is in the same way you would with a miso made of legumes. Make a quick 4% salt brine by blending 4 grams salt into 100 grams water with a handheld blender or whisk until the salt has completely dissolved. Add a little bit of the brine at a time, until the bread koji is moist enough that it can be squeezed into a ball that seems like it wants to spring back, but fails to do so. It will be quite thick and pasty, so be sure to mix it well to ensure that everything is well distributed.

Pack the mixture tightly into the fermentation vessel. Smooth and level the top, wipe the insides of the vessel clean, and sprinkle the surface with salt. Weight down and cover the breadso according to the directions outlined in the yellow peaso recipe. Let the breadso ferment at room temperature for 3 months. Check it frequently, as this miso can change character capriciously; it should develop sweet and sour twangs of umami in 7 to 8 weeks. Pack the finished breadso in airtight jars or containers and store in the refrigerator or freezer.

Suggested Uses

Breadso Soup

When breadso wears a savory hat, it brings an umami-rich warmth to everything it touches. To make breadso soup, fill a large pot with 1 kilogram chicken bones and cover with cold water. Bring it to a simmer, skimming away any impurities, then add 500 grams aromatics: chunks of leek whites, onion, carrots, celery, and garlic, along with a handful of thyme, bay, and black peppercorns. Allow the stock to simmer gently for a few hours before straining. Use a handheld blender to incorporate 150 grams breadso for each liter of stock. Season to taste with salt and finish it off by simmering 1-centimeter / ⅓-inch ribbons of savoy cabbage leaves in the soup for a couple of minutes just before serving.

Our fermented take on a classic
English sauce.

Breadso Sauce

For a variation on a classic English sauce, follow the procedure for making chicken stock, described in the Breadso Soup recipe but first roast the chicken bones at 200°C/395°F until heavily browned before adding them to the pot. Once you've made and strained the stock, reduce it in a clean pot to 20 percent of its original volume. For every 100 grams of reduction, whisk in 10 grams butter and 25 grams breadso that's been blended and passed through a tamis until it's smooth and homogeneous. To put this lip-smacking sauce to good use, fold it into sautéed morel mushrooms until they're well coated, then transfer the mixture to a shallow pan. Top with sourdough breadcrumbs and broil until crisp and golden brown. Serve while still hot and bubbling.

Breadso-Buttered Toast with Berries and Cream

Breadso is one of those ferments that can just as easily be used for dessert as it can for dinner. Try whisking together 2 parts breadso with 1 part softened butter and 1 part brown sugar. Spread the mixture over a thick slab of fresh sourdough and sear it facedown in a pan until it sizzles and caramelizes. Top it with preserved apricots or cherries and their syrup, and a dollop of freshly whipped cream.

When toasted, pumpkin seeds become aromatic and
nutty, qualities that transfer to the miso as it ferments.

Pumpkin Seed Miso

Makes about 3 kilograms

1.8 kilograms unsalted raw hulled
 pumpkin seeds
1.2 kilograms Pearl Barley Koji (page 231)
120 grams non-iodized salt, plus extra
 for sprinkling

Pumpkin seed miso was a key ingredient during our time working in Tulum, Mexico, where we used it as the backbone for our take on Yucatecan *dzikilpak*, a thick sauce or dip made from toasted pumpkin seeds. Back in Copenhagen, pumpkins are plentiful in the late summer and autumn, and pumpkin seed miso's mild richness and deep umami continues to serve us well, both at Noma and on the menu of our sister restaurant, 108.

Heat the oven to 160°C/320°F. Spread the pumpkin seeds evenly onto a few baking sheets and toast them in the oven until nutty and browned, 45 to 60 minutes. Stir and toss the seeds and rotate the baking sheets every 10 minutes or so to ensure even coloration. Allow the seeds to cool fully to room temperature.

Pulse the seeds in a food processor until they have the texture of a fine meal, then transfer to a large bowl. Place the koji in the food processor and process to break it up as well. Add the koji to the ground pumpkin seeds. Add the salt and mix everything thoroughly with gloved hands.

Make a quick 4% salt brine by blending 4 grams salt into 100 grams water with a handheld blender or whisk until the salt has completely dissolved. Add a little bit of the brine at a time until the mixture is wet enough to form a packed and firm ball in your fist without oozing out liquid or crumbling

Pumpkin Seed Miso, day 1

Day 14

Day 30

from being too dry. You'll need to add a fair bit of the brine to this miso to get it wet enough to facilitate a proper fermentation. But pumpkin seeds are a bit oily, so too much water will speed up all facets of fermentation, including the breakdown of fats into fatty acids. This can lead to rancid tastes (see the Hazelnut Miso recipe, page 317, for an explanation).

Pack the mixture tightly into the fermentation vessel. Smooth and level the top, wipe the insides of the vessel clean, and sprinkle the surface with salt. Weight the pumpkin seed miso down and cover, according to the directions outlined in the Yellow Peaso recipe (page 289). Let the miso ferment at room temperature for 3 to 4 weeks. Any longer than that, and the unwanted flavors of fatty acids will start to come through. Pack the finished miso in airtight jars or containers and store in the refrigerator or freezer.

Suggested Uses

Dzikilpak

Even though we originally developed pumpkin seed miso in chilly Copenhagen, we jumped at the opportunity to employ it on the menu of our pop-up in Tulum, Mexico. There, we were introduced to dzikilpak—a traditional Mexican salsa made from toasted pumpkin seeds. The dots practically connected themselves. The actual recipe for the sauce we made in Mexico has close to two dozen ingredients, so here we'll offer a simpler but no less delicious version.

Cut 250 grams tomatoes into chunks. Dice 1 white onion, 1 habanero, and 4 cloves garlic. Heat a medium sauté pan over medium heat and coat with vegetable oil. When the oil is beginning to smoke, add the vegetables and allow them to sear and sizzle until the liquid from the tomatoes begins to flood the pan and bubble. Transfer the pan to a 160°C/320°F oven and allow the mixture to cook down into a thick paste, about 30 minutes, stirring every 5 to 10 minutes. Transfer the mixture to a blender and add 150 grams pumpkin seed miso, a handful

Pumpkin seed miso gives Yucatecan *dzikilpak* a fermented twist.

of cilantro with the stems, and the grated zest of 2 limes. Blend the mixture until it all comes together as a smooth paste (you may need to add a bit of water to help the blender spin). Season to taste with a few spoonfuls of Beef Garum (page 373) or soy sauce. The finished sauce will be rich, spicy, thick, and fantastic with grilled seafood or in tacos of any kind.

Grilled Lettuces

If you're looking for a less involved way to put this ferment to use, try blending it with water until it's thin enough to be brushed like a lacquer. Quarter young romaine or Gem lettuce hearts and coat with the blended miso, allowing it to work its way into the cracks between the leaves. Drizzle the wedges with a bit of olive oil and season with salt before grilling facedown over a very hot grill. Char the lettuces lightly on all sides before pulling them off the grill and transferring them to a plate. Top with crumbled rye croutons and shavings of firm cheese like smoked Gruyère or Gouda.

Pumpkin Seed Miso "Ice Cream"

Pumpkin seed miso is so versatile that it has even popped up as an ice cream of sorts at Noma. Toast 200 grams pumpkin seeds in a 160°C/320°F oven until nutty and golden brown. Place the toasted seeds in a blender with 200 grams pumpkin seed miso, 750 grams water, and 140 grams good-quality honey. Blend until the mixture is very smooth, then pass it through a fine chinois. Transfer the mixture to an ice cream maker and churn according to the manufacturer's instructions. Freeze until firm, then serve with toasted coconut or almond financiers.

327

Shoyu

—

Yellow Pea Shoyu 338

Dryad's Saddle Shoyu 349

Cep Shoyu 352

Coffee Shoyu 357

The World's Most Popular Fermented Flavor

The first shoyu was, in all likelihood, a happy accident. Some Chinese cooks fermenting a batch of bean paste noticed a dark liquid pooling at the top of the container. They tasted it and were no doubt blown away by what they found. It takes a special understanding of deliciousness to look at one part of a whole and think, *Hey, this is damn good in its own right!* But that's exactly what happened. Shoyu—or soy sauce, as most of the Western world knows it—began as a by-product and grew over centuries into one of the world's most popular sauces.

Soy sauce was originally known as *jiang you*—the Chinese term for the "oil" (*you*) that pooled on top of "fermented bean paste" (*jiang*). That liquid runoff (which is actually mostly water and not oil) appears for two reasons. First, jiangs are salted ferments. The salt draws moisture out of the cooked legumes through osmosis until the salinity reaches equilibrium throughout the mixture. We see this same effect occur in lacto-ferments over the course of a couple of days, but it takes more time for the effects of osmosis to become apparent in a thicker ferment like jiang or miso.

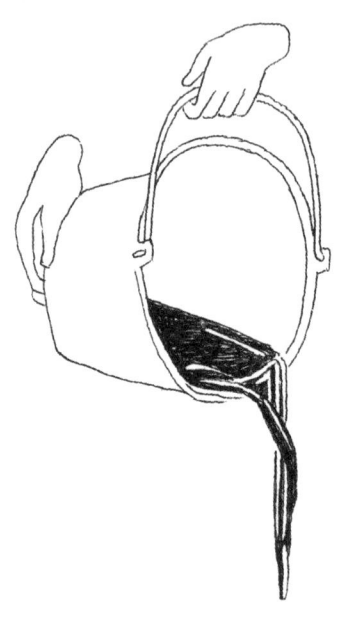

Shoyu makers learned that fermenting soybeans and wheat in brine led to greater yields.

The second factor at play is enzymatic action. Soybeans contain starches that soak up and retain water when cooked. Once the legumes are mixed with koji—that is, grains inoculated with the mold *Aspergillus oryzae*—the enzyme amylase produced by the mold works to cleave apart the starch. As starches break down into sugars, they lose their ability to gel with water, and the mixture becomes less viscous. And because jiangs were traditionally weighted down to keep air out, the liquid freed through osmosis and enzymatic action inevitably pooled and collected on top.

The terminology surrounding soy sauce can be a bit murky, so let's clarify a bit before we proceed any further. When Chinese monks brought jiangs to Japan in the sixth century, jiangs evolved into misos, and the liquid that pooled on top of miso became known as *tamari*. We'll continue to refer to it that way.

Once word got out about tamari and demand skyrocketed, Japanese artisans devised ways of producing it in greater

Many of the technical aspects of producing shoyu flow directly from, or are closely linked to, those of koji and miso production. If you're flipping through the book, this chapter will give you a thorough understanding of what shoyu is and how it's made, but we strongly encourage you to read the preceding chapters to better understand this one.

How Shoyu Gets Made

quantities that didn't involve first making miso. We refer to that product as *shoyu*, the Japanese pronunciation of the words *jiang you*. (Chinese and Japanese share many written words, but sound very different.)

The English philosopher John Locke first mentioned "saio" in a journal entry in 1679. Soon after, the German scientist and traveler Engelbert Kaempfer wrote of shoyu in his *History of Japan*, declaring Japanese "sooja" more delicious than any Chinese sauce of similar style. These Western garblings of the word *shoyu* were some of the first recorded instances of what would become the word *soya*. In a funny etymological twist, in many languages, the word for "soybeans" actually comes from these mispronunciations of "shoyu." Later publications would refer to the bean from which "soya sauce" is made as "soybeans."

The earliest Chinese jiangs blurred the line between what we would now consider the separate realms of miso, garum, and shoyu. They often contained meat or seafood, and jiang you was originally a chunky, murky liquid—far different from any soy sauce we know today. Similarly, the earliest misos of Japan were a very rough version of today's products. They contained all the same ingredients, but *hishio* (aka proto-miso) was more like a cross between a miso and a soy sauce.

Over time, hishio became more refined, eventually becoming the thick paste we know as miso today. At first, tamari remained a by-product, but as demand increased, people began making adjustments to traditional miso recipes in order to harvest more tamari. New styles of fermenting vats were manufactured with perforated troughs that led to spouts. Recipes with higher ratios of water were implemented. But it wasn't until the seventeenth century that the method for producing soy sauce evolved into the first version employed today.

We'll detail that method here, but first, let it be noted that shoyu is yet another example of a miraculous foodstuff created through the biochemical tool kit housed within *Aspergillus oryzae*. Its production bears similarities to koji and, of course,

331

miso, but there are a couple of fundamental differences in the shoyu-making process. To make miso, you combine *Aspergillus*-inoculated rice or barley with soybeans and allow them to ferment together. With shoyu, the mold is grown directly on a blend of boiled soybeans and toasted, cracked wheat. When you're making a conventional koji, you would ideally steam—rather than boil—the rice or barley so as not to oversaturate or drown the *Aspergillus*. But legumes like soybeans don't cook properly if they're steamed; they need to be submerged in water. The wheat is there to mediate the extra moisture. (This is why tamari is free of gluten, but soy sauce is not.)

Whereas in the production of something like sake, where koji is harnessed to unlock simple sugars tied up in starch, in shoyu, the aim is to use *Aspergillus* to break down vegetable proteins into the amino acids that make up soy sauce's rich medley of umami flavors. (In many soy sauce factories, *Aspergillus sojae* is the preferred strain of *Aspergillus*, as it's been bred specifically for stronger protease activity.) By growing the *Aspergillus* directly on the beans and wheat, the protease produced by the fungus gets to work directly on the proteins in those substrates.

Once inoculated, the soybean-wheat koji goes into a brine of 20% to 23% salinity. Different recipes call for different ratios of salt brine to koji, but ideally the total mixture should have a salinity of 15% to 16%. The nutrient-rich fluid is left open to the air, and the high salt content prevents unwanted microbes from taking hold. Beneficial halotolerant (salt-tolerant) microbes add a mild alcohol content and rich bouquet of complex acids to the best bottles of artisanal shoyu.

Over time, what begins as a heterogeneous soup of beans and bits of wheat (known as *moromi*—a term shared with an analogous step in sake brewing) slowly transforms into a gloopy mass with the viscous texture of baby food. As with miso, the enzymes produced by the koji slowly snip proteins into amino acids, leading soy sauce to be chock-full of glutamic acid, among other delicious organic building blocks. The enzymes work much more efficiently in a fluid medium than in a viscous one like miso.

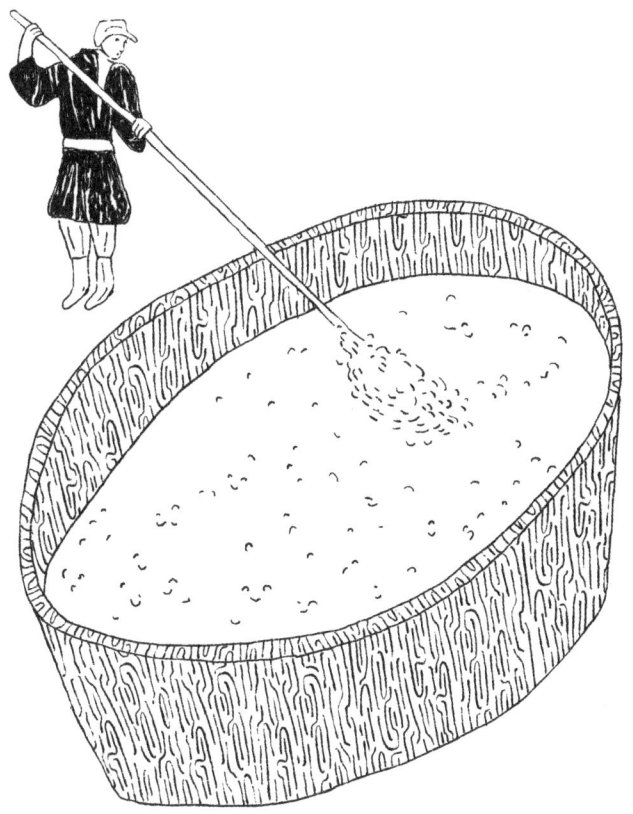

Traditionally, shoyu production would take place in massive
cedar vats called *kioke*, measuring around 2 meters across
and almost 3 meters deep. Like miso, shoyu would be started
in the winter months after the harvest, avoiding high summer
temperatures that could adversely speed up enzymatic and
microbial activity. The mixture would be stirred every day for
the first few weeks, then left to ferment for up to three years.
Once ready, the moromi would be ladled onto multiple sheets
of cloth and stacked in a large rectangular wooden press. A
huge lever would drive a wooden plate downward, pressing
the liquid out of the moromi. Once all the shoyu had been
extracted from the moromi, the remaining sediment would
be as stiff and dry as cardboard. It was often handed off to
farmers as animal feed. The strained shoyu would be left to
settle, strained once more, heat-treated, and bottled.

Shoyu,
Then and Now

Around the turn of the seventeenth century, European merchants were becoming more and more interested in Asia. England and Holland set up limited liability corporations to scour the planet with small fleets of ships in search of trade opportunities. That's how the Western world came to learn of the deliciousness of shoyu. Its addictive flavor and convenient shelf stability made it a coveted condiment.

Soy sauce fared well on the long voyage back to Europe, livening up bland food along the way. It eventually became an important ingredient in European ferments such as Worcestershire sauce, and chefs in France thought it an astonishing addition to the most traditional of dishes. Nicolas-Auguste Paillieux, a French industrialist and horticulturist in the 1800s, declared that "when a cordon-bleu cook uses [soy sauce], his cuisine is transformed, becoming much better, yet without anyone noticing that he has used moderate doses of this celebrated sauce."

While soy sauce was growing in influence in Europe and eventually North America, it was also cementing its hold over Asia. Today, the average Japanese citizen consumes ten liters of shoyu each year. There are large multinational corporations producing soy sauce for the masses, as well as artisanal outfits making small batches of otherworldly stuff. Meanwhile, soy sauce also has cousins all over Southeast Asia. Kecap manis, an Indonesian offshoot, is probably the most famous of soy sauce's kin. It's made with anise and clove and a large addition of palm sugar, and reduced until it's sweet and syrupy. In Vietnam, we find *tuong*, which hails from the north and is probably a closer relative to the original Chinese jiangs in its consistency and flavor. It's made by toasting soybeans—whereas with shoyu, you would toast the wheat and steam the beans—then lacto-fermenting them in water, and finally inoculating them with *Aspergillus*. The sauce is ground into a smooth, thick paste rather than being strained or pressed for its liquid like other soy sauces.

Other very nontraditional methods of shoyu production have cropped up as well. Acid hydrolysis is a chemical process

Kikunae Ikeda defined "umami" and founded the Ajinomoto Group.

Nordic Shoyu

developed by Julius Maggi of Switzerland's Maggi Company. It extracts free amino acids from the proteins in vegetables without any fermentation, by using hydrochloric acid and warm temperatures to break down the vegetal matter, then neutralizing the mixture with sodium carbonate. This neutralizing reaction yields salt, organic sediment called humin, and hydrolyzed vegetable protein, or HVP, in the form of a brown liquid. HVP has flavors akin to rich meat broths, thanks to the amino acid threonine.

In the early 1900s, the Japanese chemist Kikunae Ikeda used acid hydrolysis to extract the amino acids from soybeans. (Ikeda is also the man who coined the term *umami*—a portmanteau of the Japanese words *umai*, "delicious," and *mi*, "taste.") He then mixed the soybean HVP with soy sauce made from a second steeping of traditional moromi. The resulting HVP soy sauce took days, not years, to produce and cost a fraction of the price. It's never been considered as good as the real thing, but when companies first started bottling soy sauce in the United States, HVP soy (also known as chemical soy) was a popular means of production. Some manufacturers still use it. Look for "hydrolyzed soy [or vegetable] protein" on the label.

Our journey with shoyu at Noma followed a similar path to the history of soy sauce itself. When we first started making Yellow Peaso (page 289)—our version of miso—we immediately fell in love with the tamari that rose to the top of the containers. We couldn't get enough. We'd make batches of peaso just so we could have more of the liquid. Truth be told, it's probably become the single most valuable flavor component in our kitchen.

We were making and using a lot of peaso in the pursuit of tamari, as it had found its way all over the menu. But because our peasos have a relatively low salt content (4% by weight) compared to classical Japanese misos, we couldn't simply add more water to our peasos to extract more tamari the way that Japanese miso makers did a thousand years ago. If our peaso gets too wet, there's not enough salt to prevent unwanted bacteria from souring the whole batch.

Eventually, we determined that if we wanted more tamari, the only solution was to follow the logic of our predecessors and learn to make shoyu on its own. We followed classical Japanese methods, substituting Nordic ingredients for traditional ones, just as we had with our peaso. But what we ended up with was . . . shoyu. It didn't taste the same as the tamari we'd been harvesting from our peaso. It was beautiful—complex, salty, and rich—but it was also basically indistinguishable from a good-quality Japanese soy sauce.

Soy is not a flavor profile that we've mastered in this part of the world. Of course, at home we put shoyu in our chicken broth and on our eggs in the morning. We've witnessed and marveled at its culinary potential throughout Asia. We're huge fans. But at Noma, our goal is to create and nurture our diners' sense of place. We want them to draw a connection between the food they're eating and the moment and place. Whenever we've tried adding shoyu to a dish in the test kitchen, it threatens to take you out of the moment and transport you to a distant memory of a bowl of ramen in Japan or a clay pot of braised pork in Shanghai.

Some people might not notice a little drop of shoyu here and there, but others would detect it immediately. Even though our shoyu is made entirely out of local ingredients, it tastes too much like something your brain associates with other places. It's a testament to the power of association. We made Nordic shoyu, and we're proud of it, but it tastes like Japanese shoyu. For us, that's a difficult obstacle to overcome. Ultimately, we decided that rather than producing shoyu, we would "hack" our own ferments in order to generate more tamari. (You can read about our Peaso Tamari Reduction on page 298.)

All that being said, shoyu remains a remarkably versatile ingredient. You would be hard-pressed to find a kitchen—commercial or otherwise—that isn't holding a bottle of soy sauce. Marinades, sauces, stocks, and broths are all obvious destinations for shoyu, but it also brings great value to glazes, vinaigrettes, and even sweet applications such as caramel and butterscotch.

Unpacking what goes into soy sauce and learning to make it yourself is a worthwhile endeavor. Even though we've yet to figure out how to incorporate it seamlessly into our menu, we continue to experiment with shoyu and learn from each effort.

Let's quickly break down the methodology for making Noma's yellow pea shoyu:

1. Inoculate a mixture of roughly 2 parts boiled yellow peas and 1 part roasted and cracked wheat with *Aspergillus sojae* spores. Grow the koji in a fermentation chamber for 2 days.

2. Place the koji into a fermentation vessel and cover with brine. Cover the container with a breathable cloth or lid and place it in a cool room to ferment for 3 to 4 months.

3. Replace any water lost to evaporation and press the mash to harvest the liquid.

   Our primary shoyu is made with yellow peas and Konini wheat. Konini is an ancient variety, purple in color and complex in flavor once it's roasted. Konini can be a bit hard to source, but any good-quality whole-grain wheat will do. We've also had great success with rye and barley.

   This chapter also includes shoyu variations made with ingredients such as dried ceps or coffee. Once again, while we haven't found a way to integrate shoyu fully into the dining experience at Noma, we continue to experiment. Some of these recipes don't fit snugly into the definition of shoyu; some have just as much in common with garums. (We continue to call them shoyus, because unlike garums, they don't contain any animal flesh.) Our most successful concoction of this ilk is our dryad's saddle shoyu. Dryad's saddle is a type of mushroom, on which it's almost impossible to grow koji by itself, so we blend it with finished barley koji and soak it in a salt brine. It's a very different liquid from our basic shoyu—fruity, sour, salty, with a pleasant must. It's rich in umami as well, which makes it a nice, well-rounded condiment on its own, and an excellent booster to other sauces.

337

Yellow Pea Shoyu

Makes about 2 liters

600 grams dried yellow peas
600 grams wheat berries
1.9 kilograms water
365 grams non-iodized salt
Koji tane (koji spores; see Sources, page 448)

When we originally set out to make our own shoyu, we hoped that replacing the traditional input (soybeans) with a staple crop of Scandinavia (yellow peas) would yield an entirely new product. However, even though our "Nordic shoyu" is made from a very different legume, the flavor turns out to be remarkably similar to Japanese shoyu. That being said, it's absolutely worth making your own shoyu—the process is very rewarding and you'll end up with a unique yet familiar ingredient. If there's one ferment in this book that you're already adept at using, it's probably shoyu. Now you'll have an ultra-high-quality homemade version to deploy at will.

Making shoyu requires you to grow koji directly on a protein-rich substrate—in this case, peas. If you haven't attempted making rice or barley koji before, that might be a good place to begin. At the very least, first read the in-depth instructions for Pearl Barley Koji (page 231). It should also be mentioned that this recipe functions equally well with dried soybeans in place of yellow peas if you'd like to make a true-to-tradition shoyu.

Equipment Notes

You'll need a fermentation chamber (see "Building a Fermentation Chamber," page 42) and an incubating tray (made of cedar, or perforated nonreactive metal or plastic) that fits into your chamber. You'll also need a glass or plastic fermentation vessel of about 6-liter capacity, plus

338

The peas should be cooked to the point where they can be crushed between your fingers, but not so long that they become mushy.

clean cotton kitchen towels or cheesecloth and large rubber bands or a loose-fitting lid for covering the fermentation vessel. A cider press is the easiest way to harvest the shoyu, but a colander and some clean weights will suffice. We also recommend wearing sterile gloves when working with your hands, and that all your equipment be thoroughly cleaned and sanitized (see page 36).

In-Depth Instructions

Rehydrate the dried peas by soaking them in double their volume of cold water for 4 hours at room temperature.

While the peas are soaking, toast the wheat: Heat the oven to 170°C/340°F. Spread the wheat onto a large baking sheet and roast it for 1 hour, stirring every 15 minutes. The grains should be very dark, to the point where you may worry that they're approaching burnt. Intense roasting will translate to a shoyu with deeper aroma and flavor.

Remove the grains from the oven and allow them to cool to room temperature. Next, you need to crack the wheat. At Noma, we use a tabletop grain mill on the coarsest setting. If you don't have a mill, run the grains in a food processor for 45 to 60 seconds to break them up. As a last resort—and with a lot of patience—use a mortar and pestle. The aim isn't to pulverize the wheat into a fine powder, just to crack the kernel. Set the cracked wheat aside.

Now return your attention to the peas. Once they've soaked, drain the water and get them cooking in a pot filled with double their volume of cold water. Bring the water to a boil, then reduce the heat to a bare simmer, skimming any foam that accumulates on the surface. Cook the peas until they are soft enough to crush between your thumb and forefinger with light pressure, 45 to 60 minutes. You want to take care not to let the peas overcook to the point of mushiness, but even more important is not undercooking them. If the peas aren't soft enough, the koji's mycelium won't be able to penetrate their flesh and take hold.

339

Furrow the peas and wheat into rows as you would when making barley koji.

Once the peas are cooked, drain them and allow them to cool to body temperature. Once cool, weigh out 1.125 kilograms of cooked peas, place in a large bowl, and thoroughly mix them with 600 grams of the toasted and cracked wheat.

Now it's time to inoculate with the koji spores. *Koji tane* comes in two forms: either as barley or rice coated with *Aspergillus oryzae* spores, or simply as the spores themselves. It's available in various package sizes online and from home-brewing shops (see Sources, page 448). However, once you make your own koji, you can also grow your own spores for future use (see "Harvesting Your Own Spores," page 241).

Line the incubating tray (made of cedar or perforated metal or plastic) with a clean, lightly dampened towel. Spread the mixture of peas and wheat onto the towel. Using a tea strainer or powdered-sugar shaker, sift the spores over the wheat and peas. (The exact method will vary depending on the type of koji tane you've acquired; see the Pearl Barley Koji recipe, page 231, for detailed instructions.)

Set your fermentation chamber to 25°C/77°F and slide the tray inside, making sure it doesn't sit on the bottom of the chamber or too close to the heat source. Leave the chamber open slightly to allow fresh air in and heat out. It's all right for the temperature in the chamber to rise as high as 30°C/86°F, but try not to let it get any warmer than that.

After the first 24 hours, put on gloves and use your hands to break up and turn the koji, then furrow it into three rows, like mounds in a farmer's field. Place the koji back into the fermentation chamber for another 24 hours, this time increasing the heat to 29°C/84°F. By hour 48, if you're using a melanistic (non-albino) strain of *Aspergillus*, the spores will have a distinct hue and you'll see a fairly drastic color change.

Yellow Pea Shoyu, day 1

Day 14

Day 45

Day 120

Next, you need to soak the koji in brine. Bring 950 grams of the water to a boil, add the salt, and whisk to dissolve. Remove from the heat and add the rest of the water to cool down the brine.

Crumble the koji into the fermentation vessel. Shoyu is traditionally made in a cedar vat called a *kioke*. If you have access to a small one, more power to you. Otherwise, any nonreactive container with a wide mouth and straight sides will do. A food-safe bucket or glass jar of about 6-liter capacity will work fine.

Make sure the brine has cooled below 35°C/95°F, then pour it over the koji and give it a good stir with a whisk. Weigh the vessel with its contents and keep that number somewhere you'll remember—you'll need it later.

Place a sheet of plastic wrap on top of the mixture in direct contact with its surface, then cover the container with either a loose-fitting lid left slightly ajar or a breathable towel secured with a rubber band. Either way, make sure that gas can escape. Let the *moromi* (the Japanese term for the mixture we've made) ferment for 4 months in a place that's slightly cooler than normal room temperature, with normal humidity. For the first 2 weeks, give the moromi a good stir with a whisk once a day. After that, stir once a week. Whenever you stir, give the moromi a taste with a clean spoon to check on its progress. It will grow more delicious every week, with savory, roasted flavors becoming increasingly pronounced over time.

You may notice surface mold growing on your shoyu. In all likelihood, it will be the koji itself sprouting again, but it may also be kahm yeast. If you're not familiar enough to know the difference, simply skim it off.

341

Peas and wheat inoculated with *Aspergillus sojae*, hour 1

Hour 48

After 4 months, the moromi will look like a dark brown, thick, chunky applesauce. The shoyu is in there, hiding in that viscous mess, but you'll need to extract it.

First, in order to create a more fluid consistency and to balance the flavor of the shoyu, you need to calculate how much water was lost to evaporation and add it back in. This is where the starting weight of the moromi comes into play. Weigh the vessel and its contents again and subtract that weight from the starting weight. Add that amount of cold water back in.

A small cider press is the best option for extracting the shoyu from the moromi. Place the moromi in a mesh bag and press it as you would fruits for their juice. However, you can also squeeze the moromi in a cloth towel: Working in batches, scoop the moromi into a clean, sturdy cloth towel (that you're willing to sacrifice) and wring it out over a large bowl or receptacle until you're left with nothing but dry pulp. If you can't get the mash dry enough by hand, you can also set the moromi-filled towel inside a colander or the steamer insert from a pasta pot. Stack some clean weights on top and let it drain into a container until the mash is dry. If you like, you can save the leftover pulp in the freezer to backslop (see page 33) into the next batch of shoyu (around 10% of the total weight should do), giving it a healthy kick-start.

Once you've extracted all the shoyu, strain it again through a sieve lined with cheesecloth. The shoyu is quite stable and can be stored in airtight containers in the refrigerator for months, or you can freeze it for longer storage.

1. Toast the wheat until very dark.

2. Coarsely crack the wheat.

3. Cook the soaked peas until soft but not overcooked.

343

4. Mix the peas and wheat together.

5. Inoculate the mixture with koji spores and incubate for 2 days.

6. Transfer the finished pea-and-wheat koji to a fermentation vessel.

7. Cover with brine and ferment for 4 months.

8. Replace any liquid lost to evaporation.

9. Strain the mixture (aka moromi) and harvest the shoyu.

345

Suggested Uses

Shoyu-Oyster Emulsion

Oysters are a surprisingly delicious and efficient emulsifier, and the combination of ocean brine and earthy umami works quite well. Take a dozen small oysters—as fresh as you can find, of course, and definitely not preshucked—and shuck them into a blender. Add half their volume in shoyu (you can just eyeball the amount), plus the juice of half a lemon. Turn on the blender and slowly drizzle in a neutral vegetable oil until the mixture comes together like a mayonnaise.

This emulsion is superb with crunchy vegetables. Julienne half a celery root and season it liberally with salt. Let it sit, covered, for 30 minutes to pull out some of the moisture. Afterward, squeeze the julienne tightly in your fist to extract even more liquid, then dress the celery root liberally with the emulsion, a bit more lemon juice, and a fistful of chopped chives for a smashing *céleri rémoulade* that you can enjoy on its own or as a side dish.

Shoyu-Buttermilk Fried Chicken

There are many different philosophies about how to properly prepare chicken for frying. Assuming you're not married to any particular style, try this simple method: Make a marinade of equal parts buttermilk and shoyu and soak chicken legs in it overnight. The following day, remove the chicken from its bath, shaking off any excess marinade, and dredge the pieces in flour. Dip them back into the buttermilk-shoyu mixture, then again into the flour, before deep-frying in 175°C/350°F oil.

Shoyu Caramel

For a salty-sweet dessert topping, try incorporating yellow pea shoyu into caramel. In a medium pot, combine 100 grams water and 250 grams sugar. Bring the mixture to a boil, stirring occasionally to ensure that the sugar is melting evenly, especially around the edges. After 5 to 10 minutes the sugar should be completely dissolved into a light amber syrup. (A candy thermometer should read 120°C/248°F.) Add 50 grams shoyu and 200 grams heavy cream, and reduce the heat while constantly folding the contents to stop them from foaming or burning. Cook for 3 minutes before pulling off the stove and transferring to a heatproof container. You can store the caramel, covered, in the fridge, where it will thicken up quite a bit. Pull it out whenever you need to dress an apple pie, muffins, croissants, or just anything your salty-sweet tooth desires.

Whisk shoyu and cream into caramelized sugar to make shoyu caramel.

347

Wild dryad's saddle mushrooms make a shoyu with
a pronounced earthy, forest flavor.

Dryad's Saddle Shoyu

Makes about 1.5 liters

2 kilograms fresh dryad's saddle
    mushrooms
400 grams Pearl Barley Koji (page 231)
600 grams water
150 grams non-iodized salt

This is far from a straightforward shoyu, and much more of a hybrid concoction, powered by lactic fermentation. But as shown throughout the history of fermentation, sometimes blurring categories is the best way to create new ones. The forest-y flavor of this highly unconventional shoyu is a stellar seasoning for pasta dishes—whether Bolognese or *aglio e olio*—and chopped salads, seared steaks, roast birds, pan sauces, blanched broccoli, or anything, really.

Dryad's saddles (*Polyporus squamosus*) are large fan-shaped mushrooms that grow on felled trees in damp forests in the late spring, from May to June. They smell almost like watermelon rind, and the mottled brown scales on their surface evoke the plumage of fall game birds, hence their sometimes being referred to as pheasant-back mushrooms. In the company of an experienced forager, they're relatively easy to identify in the wild, and we encourage you to go out and find some. Look for specimens that are firm to the touch, and avoid anywhere worms have burrowed into the flesh. Understandably, wild mushrooms are highly dependent on the season and region, so dryad's saddle might not be available where you are. We've also had success with shoyu made from chicken of the woods and beefsteak mushrooms; but if you're not up for foraging, try hen of the woods (maitake) mushrooms instead.

349

Dryad's Saddle Shoyu, day 1

Day 7

Day 30

Brush any loose dirt and debris from the mushrooms; use a damp towel to wipe them clean if they're especially dirty. Chop them into pieces that will fit easily into a food processor, then pulse them into a coarse meal. Transfer to a nonreactive fermentation vessel. Use the food processor to break up the koji as well, then add it to the mushrooms, along with the water and salt. Stir well with a clean spoon to create a thick, uniform mash.

Place a piece of plastic wrap directly on the surface of the mash, ensuring that it reaches all the way to the edges of the vessel. Weight down the mash with some light fermentation weights or a couple of large zip-top bags partially filled with water. (Double up on the bags as a safeguard against leaks.) If the bags begin to sink into the mash, remove some water to lighten them. Cover the container, but leave the lid slightly ajar to allow gas to vent.

Let the shoyu ferment for 3 to 4 weeks at room temperature, stirring it with a clean spoon once a week. The solid matter will separate and bubble as the mixture ferments. After 4 weeks, the liquid should taste earthy, salty, and sour from lactic fermentation.

To harvest the shoyu, strain the liquid from the solids using a cider press, or by squeezing the pulp through a clean towel. Once the shoyu is extracted, strain it again through cheesecloth to ensure that all the small particles have been removed. Store the finished shoyu in the refrigerator in airtight containers or bottles. You can also freeze it for longer storage. If you intend to make future batches of this shoyu, save some of the spent pulp as well to use as backslop (see page 33) at a ratio of 10% by weight.

Suggested Use

Dryad's Saddle and Roasted Koji Sauce

If you've made dryad's saddle shoyu, then it's safe to say you've also mastered growing koji, which means you've got everything it takes to make this sauce. Crumble 250 grams koji onto a baking sheet and roast it in the oven at 160°C/320°F for 45 minutes. The sugars will brown deeply and the moldy grains will develop flavors reminiscent of chocolate. Transfer the roasted koji to a blender with 500 grams water and blend at high speed for 5 minutes. Transfer the mixture to a container to infuse at room temperature for 1 hour. Strain the mixture through a fine-mesh sieve lined with cheesecloth. Should you smell the aroma of the roasted koji water, you'll swear there's coffee in there. To complete the sauce, mix 100 grams dryad's saddle shoyu with 100 grams roasted koji water in a small saucepan and bring it up to a simmer on the stove. Using a handheld blender, emulsify 75 grams softened butter into the sauce to create a loose, buttery, salty sauce that's brilliant on lightly wilted lettuce, steamed Brussels sprouts, roasted scallops, or curls of pan-seared squid. True, it takes some heavy-duty fermentation to get to this point—you could also try it with store-bought shoyu and koji—but once you've made it, you won't look back.

351

Cep Shoyu

Makes about 2 liters

400 grams dried yellow split peas
600 grams wheat berries
Koji tane (koji spores; see Sources, page 448)
2.125 kilograms water
375 grams non-iodized salt
250 grams dried cep mushrooms

Ceps are also called porcini mushrooms or king boletes, among other regional names. Dried ceps are much easier to source than fresh ones, and you'll still get much of the earthiness you would from a shoyu made with fresh mushrooms, along with an intriguing smokiness, but less of the sour qualities of lactic fermentation.

The in-depth instructions for Yellow Pea Shoyu (page 338) also serve as a template for this shoyu recipe. We recommend you read that recipe before starting in on this one.

Soak, cook, drain, and cool the peas as directed in yellow pea shoyu. Meanwhile, toast the wheat in a 170°C/340°F oven until very deeply browned, about 1 hour, stirring frequently. Allow the wheat to cool, then use a grain mill or food processor to crack the wheat into a coarse meal.

Weigh out 700 grams of the cooled peas into a large bowl. Add the cracked wheat and mix together thoroughly. Spread the mixture onto an incubating tray lined with a lightly dampened towel and inoculate it with the koji spores. Allow to incubate at 25°C/77°F in a fermentation chamber for 1 day, then use gloved hands to turn and furrow the mixture into three rows. Increase the heat of the chamber to 29°C/84°F and incubate the koji for another 24 hours, until it begins to produce spores.

Cep Shoyu, day 1

Day 45

Day 120

Make a brine by bringing half the water to a boil, stirring in the salt, then adding the rest of the water to bring the temperature below 35°C/95°F.

In a food processor or blender, pulse the dried ceps into a powder.

Add the koji, cep powder, and brine to a nonreactive fermentation vessel and stir well with a clean spoon. Jot down the weight of the vessel and its contents and keep it somewhere you won't forget. Place a sheet of plastic wrap in direct contact with the surface of the pea and wheat mixture and cover the container, leaving the lid slightly ajar to allow gas to vent.

Let the shoyu ferment in a cool place for 4 months. Stir and taste the shoyu every day for the first 2 weeks and once a week after that. Once it's finished fermenting, weigh the vessel and its contents again, and subtract that weight from your starting weight to calculate how much moisture has been lost over time. Add that amount of cold water back in.

To harvest the shoyu, strain the liquid from the solids using a cider press or by squeezing the pulp through a clean towel. Once the shoyu is extracted, strain it again through a sieve lined with cheesecloth to ensure that all the small particles have been strained out. Store the finished shoyu in the refrigerator in airtight containers or bottles. You can also freeze it for longer storage. If you intend to make future batches of this shoyu, save some of the spent pulp to use as backslop (see page 33) at a ratio of 10% by weight.

Suggested Uses

Cep Shoyu Beurre Blanc

As first demonstrated by the late, great Alain Senderens, soy sauce is right at home in a French beurre blanc. Add about 150 milliliters white wine to a saucepan and reduce by two-thirds; maybe add some minced shallots along with black peppercorns. Reduce the heat to a bare simmer. Whisk in 100 grams cubed cold butter—yes, it's a lot, but it's worth it—one at a time. Don't let the sauce boil or it will break; heat to just keep the mixture warm. Once all the butter is incorporated, pull the sauce off the heat and park it beside the stove to keep warm. Right before serving, whisk the sauce vigorously and season with about 50 milliliters cep shoyu.

This sauce is fantastic with steamed or pan-seared fish, or steamed greens. Cook chopped kale leaves with just a bit of water until wilted, then season with salt. Toss in some ramson capers or slivers of tart berries like green gooseberries. Remove from the heat and coat the leaves with beurre blanc. Serve in a bowl with croutons.

Cep Shoyu–Glazed Ceps

Cep shoyu presents an opportunity to double down on the flavor of cooked mushrooms. Halve a few fresh ceps lengthwise and score the cut sides in a crosshatch pattern. In a hot pan, heat enough clarified butter to coat the surface and sear the ceps, scored-side down. As the ceps begin to brown, reduce the heat and add a knob of butter, a crushed clove of garlic, and a sprig of thyme. Let it foam and sizzle, then flip the ceps and baste with butter until they're just about cooked through. Drain off the fat and return the pan to the heat. Add a healthy splash of cep shoyu and let it bubble and reduce. Add a fresh spoonful of butter, swirling the pan to help bind the liquids into a glaze and lacquer the ceps. Remove them from the pan; season with a few drops of lemon juice or, better yet, lacto cep juice (page 83).

Pan-roast fresh ceps with cep shoyu
to double down on both the umami and
mushroom flavors.

355

Coffee, already a fermented product,
adds depth and complexity when fermented
again as shoyu.

356

Coffee Shoyu

Makes about 1 liter

800 grams Pearl Barley Koji (page 231)
200 grams leftover coffee grounds or
    100 grams freshly ground coffee
1 kilogram water
80 grams non-iodized salt

The first thing you'll notice about this shoyu recipe is that it doesn't include any legumes—no soybeans, no yellow peas. We initially developed this recipe as an interesting way to utilize leftover coffee grounds, and to be honest, it doesn't fit neatly into the shoyu box or any of the ferments in this book. Rather than fermenting at room temperature, we keep coffee shoyu in a fermentation chamber as we do with our garums, to bring out more roasted flavors and speed up enzymatic action. However, it's modeled after shoyu, so we call it coffee shoyu.

Pulse the koji in a food processor until it's thoroughly broken up into small grains. Transfer to a large bowl and add the coffee grounds, water, and salt.

Transfer the mixture to a fermentation vessel (a food-safe plastic bucket or 4-liter glass jar with a lid). You can also ferment the mixture directly in the bowl of a rice cooker set to "keep warm." Because we ferment this shoyu at a high temperature, 60°C/140°F, evaporation is more of an issue than it is with other shoyus. To prevent moisture loss, double-wrap the fermentation vessel with plastic, even if it has a lid. Put the vessel into your fermentation chamber; if using a rice cooker, wrap the lid of the cooker in plastic wrap as well.

Let the coffee shoyu ferment for 4 weeks, stirring and tasting once a week. You should find the finished shoyu bittersweet, with a flight of roasted fruit flavors. Once you're satisfied,

357

Coffee Shoyu, day 1

Day 7

Day 28

strain the shoyu through a fine-mesh sieve and then re-strain through a sieve lined with cheesecloth. The finished shoyu can be stored in the refrigerator in airtight containers or bottles for months. You can also freeze it for longer storage.

Suggested Uses

Fish Glaze

You can reduce coffee shoyu in a pan—slowly and carefully, lest it burn—into a delicious syrup. A teaspoon of this syrup added to the pan 20 seconds before you finish pan-frying a few fillets of fish will coat them with a profound sweet-saltiness.

Coffee Shoyu Butterscotch

Here's a bit of a quirky suggestion for coffee shoyu: Mix it into butterscotch. You've probably tasted salted caramel or salty butterscotch. In this case, the saltiness comes with the added complexity of fermentation. Plus, it's not terribly complicated to make your own butterscotch, and it's really, *really* good. In a medium pot over medium-low heat, melt 60 grams butter, then add 100 grams dark brown sugar, 125 grams heavy cream, and 60 grams coffee shoyu. Boil the mixture for 4 to 5 minutes, then add the seeds of half a vanilla bean. Stir and remove from the heat. Allow it to cool and refrigerate, covered, to use as a sauce for cakes, pies, or any confection of your choosing.

Overnight Chicken Broth

This is a super-simple way to start your day with a wholesome, straightforward meal. Place the carcass of a whole roast chicken into a pot, cover it with water, add a few aromatics, and let it bubble away over low heat for an evening. Before you go to bed, turn off the heat, cover the pot with a lid, and leave it on the stove until morning. The next morning, strain the broth and season with coffee shoyu—or any shoyu, really. Add noodles, rice, or vegetables for an even better start to the day.

Slowly reduce coffee shoyu into a syrupy glaze,
then use it to lacquer fillets of sole.

359

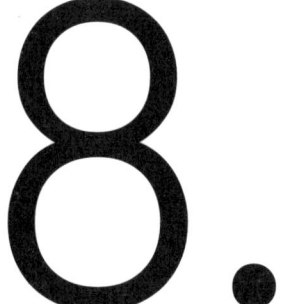

Garum

—

Beef Garum 373

Rose and Shrimp Garum 381

Squid Garum 385

Roasted Chicken Wing Garum 389

Grasshopper Garum 393

Bee Pollen Garum 397

Yeast Garum 400

Got to Have
the Funk

Thomas Frebel, former head of Noma's
test kitchen, first suggested we make garum
from meat rather than fish.

A little bit goes a long way, when it comes to certain things: honesty, kindness, axle grease . . . and fish sauce.

Garums—the larger family to which fish sauce belongs— are a largely forgotten ingredient in the West. Once a mainstay of European cuisine, they've all but disappeared from the recipes of today. In its purest form, garum is a chunky blend of fish, salt, and water that's allowed to break down and putrefy (in a controlled manner, of course). We use the term *garum* somewhat more broadly at Noma, and expand the ingredients to include a lot more than just fish.

Thomas Frebel, the former head of our test kitchen, was the first to suggest we try making garums with meat rather than fish. At the time, we'd been struggling with how we could make ancient traditions like garum feel new and distinctly ours. Thomas's suggestion proved to be a brilliant one.

Garums are relatively easy to produce, and it turns out that the process works just as well with meat as it does with fish. We also found that if you add koji to the equation, you can reduce the time it takes to make garum by more than half. (Without the koji, many of the garums we make are technically not products of fermentation, but rather autolysis. More on this later.)

After a good amount of trial and error, we can say with confidence that the way that we make garum at Noma— fermenting animal protein in warm conditions with salt, water, and koji—is a novel twist on the traditional methodology. The resulting garums are quickly becoming among the handiest ingredients in our arsenal. They don't play a starring role, but they're there under the surface, imbuing dishes with an intangible magic, focusing and enlivening natural flavors. If you'll forgive us for inventing words, they give things *intricity*, which is the portmanteau we've come up with to capture the intensity and electricity that ferments bring to cooking. There's simply no other way we can think of to describe the effect of adding a teaspoon of squid garum to a big pot of steamed potatoes tossed with melted butter and a fistful of chopped parsley. The flavors come to a point—they have depth and umami, and taste like enhanced versions of themselves.

The most exciting thing is that we're only beginning to understand the potential of garums. Rather than reaching for a pinch of salt in a recipe, we'll sometimes kill two birds with one stone by using garum to bring both salinity and umami. And at a restaurant like Noma, where meat doesn't play a huge role in the menu for most of the year, garums can give you the satisfaction of having eaten a piece of beef or chicken without the heaviness. When we do serve meat, we'll often use a corresponding garum to up the intensity, whether that means a few drops of beef garum on strips of raw beef, or squid garum to enhance pieces of squid cured between sheets of kelp.

In a way, garums have allowed us to reverse the role of animals and vegetables at Noma, so meat becomes the seasoning and vegetables are the stars. A splash of garum can elevate an unassuming cabbage leaf to a satisfying and memorable bite. It's really how we should all be eating, anyway. Before meat was a commodity, it was a luxury. When you could get your hands on it, you had to make it last. The earliest Chinese *jiang*s were mixtures of meat, soybeans, and *Aspergillus*, and they occupied a similar place as garums in the local cuisine. And in Scandinavia, people have been curing herring for centuries, and using the liquid runoff as a seasoning. They didn't call it

A little garum goes a long way.

363

"garum," but they recognized its usefulness. It's all about stretching the resources you've got—a practice that often leads to delicious innovations.

Carthaginian Fish Sauce

Garum's story begins in North Africa 2,500 years ago, when the walled Phoenician metropolis of Carthage was a booming port in what is present-day Tunisia. Within the city walls, surpluses of fish harvested from the flourishing waters of the Mediterranean Sea—tuna, mackerel, anchovies, sardines—were sliced finely, scales, heads, guts, and all, then layered with salt in limestone vats and left to ferment. Nets draped over the vats would keep larger animals and flies out. The heat of the sun would effectively cook the fish, while the salinity would act as a safeguard against the propagation of harmful microbes. Most important, the guts of the fish contained enzymes that fueled the transformation from vat of fish parts to potent seasoning.

The Carthaginians enjoyed a reign over the Mediterranean for close to five hundred years, until the city fell to the Roman Empire in the Second Punic Wars. To the victor go the spoils, and as Carthage changed hands, so did its culinary practices. Though the product originated in North Africa, the Romans are the ones credited with the dissemination of *garum*—a Latin word derived from the name of a specific fish species. Sicily, with its proximity to Carthage, was likely where the gospel of fish sauce first spread, and it served as the center of garum production in the ancient Roman Empire.

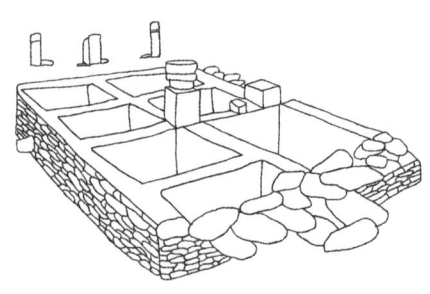

Ancient Carthaginian garum factories were carved out of limestone in ports along the Mediterranean Sea.

In much the same way that nuoc mam is used in Vietnamese cuisine as both a dipping sauce and a seasoning, garums would have been served at the table but also kept in the kitchen to be used with wine as a cooking sauce called *oenogarum*. Garums served the Roman army well, too. Soldiers could carry the concentrated salty liquid in a flask and dilute it while in the field. After the Third Punic Wars and the Roman annexation of Iberia, garums spread west. In southern Spain, ruins of garum factories carved out of limestone are still standing to this day.

As garum grew, specialized classifications emerged. The sediment left over after straining a batch of garum was known as *allec*. It was deemed undesirable by elites and left to the commoners. *Muria* was garum made from fish that were gutted and had their heads removed, which would have rendered the finished sauce less pungent. *Haimation*, a fermented product that consisted of nothing but the guts and blood of fish, was made from the by-products of fisheries. Its dark color also led it to be dubbed "black garum." *Liquamen* was once a term distinct from garum, as it was used concurrently in early Roman times, though it isn't entirely clear what the difference was. Some believe it to be a second steeping of the allec in an attempt to extract more yield from the fermented fish. Others explain it as garum made specifically with whole fish, whereas *garum* is an umbrella term for the larger family of related sauces.

What's even less clear is why garum fell out of fashion in the West. The last vestige of garum in Europe is a rare Italian sauce called *colatura di alici*, traditionally made in the small fishing village of Cetara. Its recipe was recovered by monks in the Middle Ages from much older Roman texts. Meanwhile, fish sauce remains a bedrock of Southeast Asian cuisine and an ingredient that many of us are far more familiar with. The process of making it is remarkably similar. Anchovies fished from the Gulf of Thailand are placed into large wooden vats, layered with salt in a ratio of two or three parts fish to one part salt. The salted fish are pressed beneath a bamboo mat weighted down with rocks and left in the tropical sun for 9 to 12 months, before being pressed and strained for their juices. Most of the fish sauce you've tasted in your life is made this way.

The curious thing about Asian fish sauces is that there aren't many references to them in historical documents from the region before the seventh century. Cultural exchange between the Roman Empire and Asia was established long before then. Given garum's value to ancient Romans and its portability, it's tempting to draw a connection between Thai fish sauce and garum, as opposed to assuming both products were developed independently. It's fun to imagine a direct line between the vastly disparate cooking styles of Southeast Asia and the Mediterranean, but we'll leave that for more qualified parties to decide.

Fish Digesting Fish

The fact that fish sauce stinks is nothing new to anyone who's ever cooked with it. But fish sauce doesn't actually smell *fishy*—at least, not if it's made well. Fishiness is a result of fish flesh and fats being spoiled by bacteria. If the fish in a garum isn't fresh, the finished garum will suffer for it. Fish innards—the main catalysts responsible for making garum—have a pungency that's quite different, earthier and less offensive than rotting fish.

The traditional method of making garum combines wild fermentation with autolysis. An autolytic process is one in which an organism's tissue or cells are broken down by enzymes produced by the organism itself. In other words, to make garum, you turn an animal's normal digestive process on itself.

The flesh of all animals contains proteolytic (protein-dismantling) enzymes that contribute to autolysis. If you're wondering why you're not digesting yourself right now, it's because those enzymes are present in extremely small quantities, and in an organism's healthy cells, they are sequestered within an organelle known as a lysosome. But once an animal dies, its enzymes act upon its flesh indiscriminately. Take dry-aged meat, for example: When a cut of beef is left on a shelf in the fridge, the enzymes it contains will slowly break down its connective tissue and muscle, tenderizing the meat and making it more delicious as the proteins are snipped into their constituent amino acids.

Making garum is essentially the same thing as dry-aging beef, only wetter, faster, and more intense. Rather than harnessing the enzymes in an animal's flesh, garums depend on the enzymes in the gastrointestinal tract, which are even more concentrated and potent. An essential part of making garums the traditional way is to chop up the whole fish—guts, flesh, and all. As the fish sits in vats with salt, the digestive juices (both stomach acids and intestinal enzymes alike) come into contact with the flesh of the fish they're normally kept separate from. The juices go to work on the fish flesh, breaking down proteins into their constituent amino acids, and fats

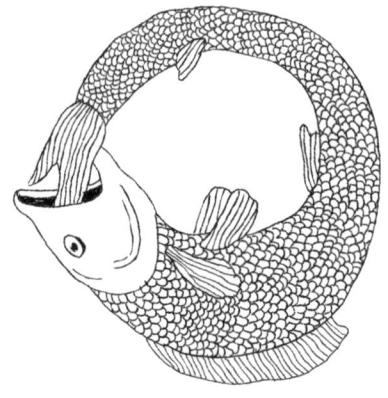

Autolysis is the term used to describe an organism digesting itself.

into fatty acids. The salt does double duty, expediting autolysis while simultaneously safeguarding the mixture against harmful microbes. That said, there *are* a handful of halotolerant (salt-tolerant) microbes that live in a mash of saline fish, adding to the garum's bouquet of volatile aromas in much the same way that shoyu's community of beneficial microbes does.

Enzymes need to be suspended in a liquid medium to function efficiently; otherwise they won't be able to float from one protein chain to another, snipping them into amino acids as they go. Salt draws moisture out of the fish into its surrounding environment through osmosis, creating a soupy environment for enzymes to travel through. As the fish's muscles break down, the salt has an even easier time drawing out more water. The process snowballs, liquefying solid fish into garum. (Heat also precipitates enzymatic reactions, which explains why garums were traditionally fermented under the hot Mediterranean sun. Temperatures hovered around 30°C/86°F in the summer in ancient Carthage; a garum would near completion in 6 to 9 months in that kind of heat.)

Salt/Water

Salt's other role in a vat of garum is to prevent spoilage. As mentioned many times in this book, there are plenty of halotolerant spoilage bacteria that have no problem living in mildly saline environments. But there is a limit, and garum's salinity sits over that limit. Highly salted solutions safeguard against spoilage via two mechanisms—osmosis, which you've already read about, and another property called water activity, which extends to all types of fermentation.

Water activity is not a measure of how much water is within a product, but rather how tightly bound the water is to that product. It's a measure of how much water vapor a sample gives off, expressed as a ratio. Distilled water has a water activity of 1, while a totally dry substance—for instance, sand that's been baked in an oven so that any water inside it has evaporated—has a water activity of 0. Dried fruit has a water activity of about 0.6; raw meat, about 0.99. Most bacteria need an environment with a water activity above 0.9 to grow; fungi, above 0.7. (Freezing, by way of locking water molecules

When surrounded by a salt solution, the water in cells travels outward to areas of higher ionic concentration. As a result, the cell shrivels and dies.

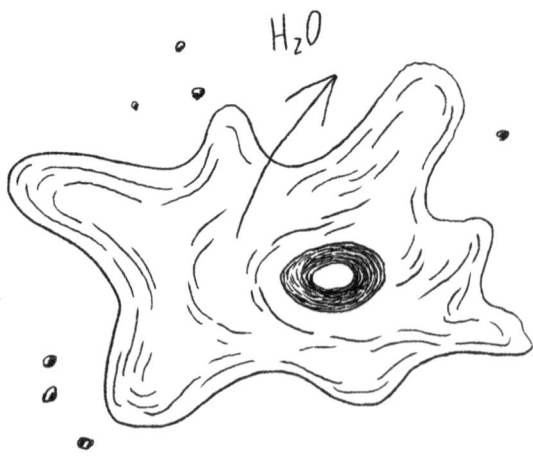

H_2O

into a rigid lattice, also effectively lowers water activity, and is the reason it's such an effective method of preservation.)

In a batch of garum, salt lowers the water activity of a mixture by binding to individual water molecules, effectively removing them from the solution. Because the water molecules are sequestered by salt ions, they are unavailable to the microbe's normal life processes. This works in concert with osmosis, which acts the same way on microbial cells as it does on meat or fish. The salt draws water out of the microbe's cells, collapsing them so they shrivel and die. This mechanism staves off spoilage in all sufficiently salted products, not just in garums, but also aged cheeses, cured meats, misos, shoyus, and lactic ferments.

Better Living Through Koji

The flavor molecule most responsible for garum's delicious-ness is glutamic acid. Glutamic acid is an amino acid that is present in almost all proteins. In its free form (just hanging out, not part of a protein chain), it can be found in especially high concentrations in meats, cheeses, tomatoes, seaweed, and wheat. When the proteolytic enzymes in a vat of garum cleave apart the proteins in fish or meat, it frees molecules of glutamic acid, which then give up a free positive charge to become glutamate. Glutamate, in turn, binds to mineral ions like sodium to form monosodium glutamate (MSG).

Aside from being a well-known powdered food additive, MSG is naturally responsible for some of the world's most delicious foods, from ramen to risotto. It registers on the tongue not as a flavor, per se, but as the sensation of umami—the "fifth element" of taste, often thought of as the essence of savoriness, first postulated in the early 1900s by the Japanese chemist Kikunae Ikeda. Perhaps the best descriptor of its flavor is "moreish"—as in, when you taste it, you want more. Glutamates can even evoke the physiological reaction of salivation, literally making your mouth water.

We're primed for umami from birth—human breast milk contains ten times as much free glutamate as cow's milk. During breastfeeding, the glutamate content in milk rises steadily as infants nurse, to the point where it can account for as much as 50 percent of the total free amino acids. We even have glutamate receptors in our gut that signal our brains as we begin eating something rich in umami: Our appetite immediately increases, but we feel sated sooner and for longer than when we eat a low-umami meal. We're hardwired to find umami satisfying, and we seek it out as a result.

While the most pronounced feature of garums and fish sauces is the powerful funk generated by autolysis and fermentation, the smell is actually misleading. The stinkiness is something you come to appreciate, but the glutamates are the foundation of garum's appeal, elevating all that they touch. Now, say you wanted to mitigate garum's aroma while maintaining its complexity and glutamic content. You could omit the guts responsible for autolysis, but you'd need some other tool to break down protein into amino acids. And in koji, once again, we find a friend.

Glutamate ($C_5H_8NO_4$) is the delicious taste of umami in molecular form.

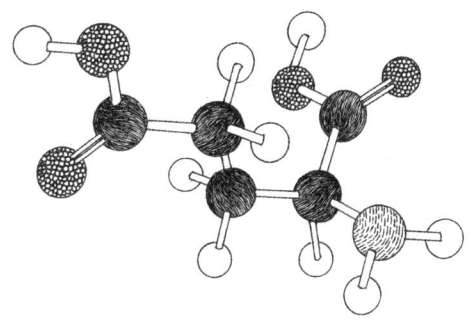

Garum

370

Steamed crab is dressed with a sauce
made from egg yolks cured in kangaroo
and beef garums.

Koji produces enzymes called proteases, which we use at Noma to break down the proteins in beef, squid, mackerel, clams, and other protein sources. Put simply, koji steps in and does the job of the digestive enzymes in the guts of fish, yielding a finished product with as much umami as the traditional method, but a far more pleasant smell.

In order to produce a faster garum, we also ferment ours in a room held at 60°C/140°F. While this temperature precludes microbial activity, it accelerates enzymatic activity to a maximum while simultaneously facilitating Maillard reactions that imbue the sauce with the flavor of roasted meat. At this temperature, we can typically go from a bucket of meat to fully finished garum in 10 to 12 weeks. You'll notice distinctive changes in the product as those weeks go by. At first, it tastes like a murky stock, but after the first week or so, the enzymatic action takes off and you can sense the umami building. After about a month, more caramelized flavors step to the front. By the end, everything sits in delicious harmony.

It's possible that you find something inherently worrisome about placing raw flesh into salt water with moldy grains and letting them sit for months, but rest assured, garums are by far the most precise and safe ferments we make at Noma. The high salt content (about 12% by weight), coupled with the high temperature, creates an environment that nearly all food-borne pathogens can't tolerate.

Meanwhile, we're continually trying to dissect garum and rebuild it in different ways. We've experimented with a variation that omits water, and ended up with a much thicker mixture that resembles Thai shrimp paste. We've made garums with protein-rich plant matter, like peas, that work well but shouldn't be left in the heat for as long as animal protein. We've also tried garums with more unconventional sources of protein, like bee pollen, grasshoppers, moth larvae, and pig's blood. There's plenty more to explore in this vein. For the adventurous spirits out there, take a look at pineapple and papaya. They're both fruits heavy with proteolytic enzymes, and you might be able to conjure up a tropical garum. We don't get many pineapples here in Denmark, but it's an idea.

371

Garum was traditionally a sauce made from decomposing fish, but at Noma we begin with ground beef.

Beef Garum

Makes 1.5 liters

1 kilogram freshly ground lean beef
225 grams Pearl Barley Koji (page 231)
800 grams water
240 grams non-iodized salt

Beef garum truly took off at Noma around the time we were serving beef ribs on the menu, and there were a lot of beef scraps around. With any ferment, your ingredients must be fresh and pristine to avoid the threat of spoilage or mold. This holds especially true when it comes to meat- or seafood-based ferments, and it remains the case even if you're using meat scraps that might otherwise be thrown away. If it's not fresh enough to eat, it's not fresh enough to ferment. For this recipe, you can grind the beef yourself or ask a butcher to do it for you, but avoid ground beef that wasn't ground fresh the same day you start your garum.

Finally, *Aspergillus oryzae*-inoculated koji works fine for this process (it's what we use at the restaurant), but *Aspergillus sojae*, mentioned in the shoyu chapter (page 332), is especially well tailored for garum fermentation. *A. sojae* produces higher levels of protease than other strains, so it more effectively breaks down the beef, producing higher levels of glutamate and thus more umami.

Equipment Notes

Our garum is fermented at 60°C/140°F, so it will require a fermentation chamber (see "Building a Fermentation Chamber," page 42) or large-capacity electric rice cooker or slow cooker. (For those who are devoted to historical authenticity, see

Beef Garum, day 1

Day 7

Day 30

"Classical Garum," page 378, for instructions on fermenting garum at room temperature.) Otherwise, garum only needs a food-safe container to hold it. We use 30-liter brewer's buckets for the large quantities we produce at the restaurant, but for this recipe, you only need a 3-liter container. Glass jars and classic ceramic fermenting crocks work great, too.

In-Depth Instructions

Place the beef, koji, water, and salt into the fermentation vessel of your choosing. Use a handheld blender or gloved hands to mix everything thoroughly. Scrape down the inner sides of the container, then cover the surface of the garum with a sheet of plastic wrap, making sure it comes into contact with the liquid and the sides of the container to create an airtight barrier. Cover the container with a lid, and screw it on slightly less than completely tight if it's a screw cap or leave it slightly ajar in one corner if it's a snap lid to allow a bit of gas to escape.

Move the garum into the fermentation chamber and set the temperature to 60°C/140°F. If you're using an electric rice cooker or slow cooker, use a sushi mat or wire rack as a buffer between the bottom of the bowl and the fermentation vessel; set it to "keep warm." (If the bowl of your slow cooker or rice cooker has a capacity that's close to the total volume of the garum mixture, you can forgo the fermentation vessel and ferment the garum directly in the bowl of the cooker.)

Allow the garum to ferment for 10 weeks. It will separate as it ages, with the ground meat floating to the top like a raft and the liquid sitting at the bottom. The salt and heat should keep malevolent microbes away, but the beef fat will begin to degrade into free fatty acids that can take on musty flavors, which can come across as rancid. As a countermeasure, several times during the first week remove the lid and plastic wrap and use a clean spoon or ladle to scoop off as much fat as possible. Stir the garum and replace the coverings. After the first week, you'll only need to skim and stir the mixture weekly. After 10 weeks, the beef garum should be dark brown with a roasted, nutty aroma and deep, meaty, rich flavor.

Day 75

Strain the garum through a fine-mesh sieve, pressing out as much liquid as possible without allowing any solids to pass through. Then strain the liquid again through a sieve lined with cheesecloth. The solids can be reserved and added to misos or used as a seasoning.

If there's any fat resting on top of the liquid, skim it off using a ladle or spoon. Pour the liquid into bottles or another covered container. The garum is very stable and will keep well in the fridge for at least a month. You can also freeze it for longer storage without any negative effects, but note that because of the high salt content, it probably won't freeze completely solid.

375

1. Freshly ground lean beef, water, koji, and salt.

2. Mix all the ingredients in the fermentation vessel by hand or with a handheld blender.

3. Cover the garum with plastic wrap and a lid and ferment at 60°C/140°F.

376

4. Skim the fat from the garum several times during the first week, stirring each time and then covering the liquid and vessel again.

5. Ferment for 9 weeks more, skimming and stirring once a week.

6. Strain the garum and store, covered, in the refrigerator or freezer.

377

The in-depth recipe describes the way we produce garum at Noma, but it's not the only way to do it. Ancient Carthaginians and Romans (along with most of the producers in Southeast Asia today) fermented their garums at ambient temperature. They also depended on proteolytic enzymes within the guts of fish, rather than the power of koji. Here we'll outline two traditional methods.

To make beef garum at room temperature (without a fermentation chamber): Increase the salt to 365 grams (18% by weight) to prevent spoilage. Place in a food-safe glass, ceramic, or plastic container and top the surface with plastic wrap. Ferment the garum for 8 to 9 months, covered but not sealed airtight. Stir often for even fermentation and mold prevention. If you see mold on the surface, remove it immediately. The finished liquid will be reddish to amber colored and smell slightly musty and sweaty, but will taste extremely clean, with deep layers of umami and a subtle beefiness. Use this adjustment on any garum in this chapter.

To make garum without koji, you'll need another source of protease. To avoid *E. coli* contamination, don't use the innards of a cow. Instead, opt to make a fish garum, using whole mackerel, smelt, or sardines (with guts). Cut the fish into chunks—heads, fins, meat, bones, guts, and all—and blend in a food processor or blender. For fermentation at 60°C/140°F, add 12% of the fish's weight in salt. For room-temperature fermentation, add 18%. This method is much closer to a traditional garum—it will smell much stronger, but it will still be just as tasty.

Suggested Uses

Egg Yolk Sauce

Beef garum forms the backbone of one of the most popular dishes we've ever served at Noma—crab with egg yolk—and it's only fitting that we share the secret here. To us, this simple combination is the perfect sauce, and its uses are myriad. Start by separating 4 eggs and placing the yolks in a bowl. (If you're squeamish about raw egg yolks, you can soft-boil the eggs before harvesting the yolks.) Whisk in 15 grams strained beef garum. That's it. Boil a head of cauliflower and cut it into small florets. Dress each serving with a couple of generous spoonfuls of the egg yolk sauce, and season with a little extra salt, chopped parsley, and lots of freshly cracked black pepper. Or dress a roasted sweet potato with a dollop of butter, a teaspoon of honey, and chopped chives and serve with the egg yolk sauce for a hearty (mostly) vegetarian meal. Or, you could just as profitably serve the same sauce as an accompaniment to steak with some greens on the side. A handful of tender greens and a few good crunchy-soft croutons dressed in egg yolk sauce would make a full lunch. Finally, try serving it as a dip for a plate of spicy, crunchy radishes at the height of the summer, paired with a glass of Champagne or beer.

Pasta with Egg Yolk Sauce

One more wonderful use for the egg yolk sauce is a quick pasta. Whisk 2 heaping spoonfuls of finely grated Parmigiano into the sauce. Cook 225 grams of your favorite pasta shape until al dente. While the pasta is still steaming hot, mix in the egg yolk sauce. At this point, it's ready to eat, but you could also give it a generous dose of freshly cracked black pepper, or add a heap of chopped tomatoes or fresh basil. It's a perfect weeknight dinner that kids will love, too.

Burgers, Broths, and Beyond

Hamburgers are greatly improved by incorporating a spoonful of beef garum into the meat before shaping it into patties. (Or you could add the leftover solids from the garum to great effect.) Really, you should think of garum as a meatier version of soy sauce. Just about any broth or soup you'd ever cook could benefit from a shot of beef garum. The same goes for stir-fries and sauces.

Beef Garum Emulsion

As mentioned, the leftover solids are still chock-full of flavor and need not be thrown away. In a frying pan over medium heat, slowly render and pan-roast 250 grams of the strained beef garum solids. As the solids caramelize and crisp up, they'll begin to release fat. Keep cooking until they've crisped up like lardons, without rendering all the way through; then, while still warm, transfer them to a bar blender and begin to blend on high speed. Slowly drizzle in an equal amount of neutral oil, as if making mayonnaise. The mixture will thicken and emulsify. Finish by brightening the mixture with lemon juice, lacto cep juice (page 83), or black garlic vinegar (page 206). Use the emulsion as a sublime dressing or dip for cooked or raw vegetables, topped with a grating of fresh horseradish.

Egg yolks curing in beef garum.

379

In this garum, the pungency of fermented shrimp is offset by the floral perfume of wild roses.

380

Rose and Shrimp Garum

Makes about 3 liters

1 kilogram fresh head-on, shell-on
    small northern shrimp
1 kilogram water
500 grams wild rose petals
450 grams non-iodized salt

This recipe is much closer to a classical garum, as it uses no koji, meaning the shrimp are allowed to truly autolyze as their guts get blended along with everything else. The shrimp we use for this garum are tiny and easily blended whole. For the most part, this ferment also takes place at ambient temperature, rather than in a hot fermentation chamber, in order to retain the rose's floral notes. The flower's sweet aroma is a perfect foil to the fishy funk of the fermented shrimp.

The in-depth instructions for Beef Garum (page 373) serve as a template for all the garum recipes in this chapter. We recommend you read that recipe before starting in on this one.

If you can't source small northern shrimp, other locally available shrimp varieties will work; just be sure to use wild, not farmed, shrimp.

In a food processor or blender, blend all the ingredients together until you have a smooth paste. Transfer to the bowl of a rice cooker or slow cooker, cover, and set to "keep warm." If your appliance lacks a rubber seal and a latch, wrap the whole appliance in plastic wrap to prevent moisture loss.

Allow the mixture to ferment in the cooker for 24 hours, then transfer to the fermentation vessel of your choice (be sure it has a minimum capacity of 3 liters). Scrape down the inner sides of the container with gloved hands or a rubber spatula

Rose and Shrimp Garum, day 1

Day 7

Day 75

and lay a sheet of plastic wrap directly on the surface of the liquid. Cover the vessel loosely with a lid, leaving at least a corner slightly ajar. Allow the garum to ferment at room temperature for 2 to 3 months, stirring once a week. This will be a deeply stinky ferment, but stinky in a good way—the way truffles are off-putting and enticing at the same time.

To harvest, strain the garum through a fine-mesh sieve lined with cheesecloth. Reserve the solids to use as a seasoning paste of sorts. You may want to blend it even finer and then pass it through a fine-mesh sieve to improve the texture. Use it as you would a Thai shrimp paste to start curries, adding it to the pot as you're sweating your aromatics; or add a tiny pinch to spicy dipping sauces made of rice vinegar, shoyu, and chili oil.

Pour the garum into bottles or another covered container. The garum is very stable and will keep well in the fridge for months. You can also freeze it for longer storage without any negative effects, but note that because of the high salt content, it probably won't freeze completely solid.

Suggested Uses

Seafood Accompaniment

In the best possible way, rose and shrimp garum tastes like a greatly reduced shellfish broth with a tiny bit too much salt. Use it the same way you would a fish sauce, in dishes where you're striving to add character but a little less funkiness than something like squid garum, which can sometimes have notes of strong cheese. We love this garum as a complement for seafood, combined with nothing more than an equal portion of good olive oil. It's great on raw, steamed, or grilled shrimp. Or the next time you're steaming clams or making clam chowder, add a teaspoon of shrimp garum to each serving in place of salt.

Butternut Squash Soup

In the early fall, when the weather turns and winter gourds are at their best, butternut squash soup starts to show up in restaurants and home kitchens around the world. It's delicious, but predictable. Take it in a completely different direction by adding rose and shrimp garum, which will simultaneously echo and contrast the flavor profiles at work. Simmer chunks of peeled and seeded squash in enough vegetable or chicken stock to cover. Once the squash is completely softened, puree it in a blender with the stock. With the blender still whirring, add a teaspoon of garum per serving; you'll immediately be hit with an inimitable aroma. Finish the soup with a dollop of whipped crème fraîche and a light grating of lime zest.

North Sea squid produce an intensely aromatic and flavorful garum.

384

Squid Garum

Makes 2 liters

1 kilogram whole squid, including
    innards and ink, but with the beak
    and cuttlebone removed
225 grams Pearl Barley Koji (page 231)
800 grams water
240 grams non-iodized salt

A recurring theme in our fermentation projects is finding a second life for waste. At one point, we were serving the tender portions of large North Sea squid on the menu, leaving us with a lot of guts, tentacles, and tough ends. This garum, made from those leftovers, was one of the first we produced at Noma, and it's still among the most successful things to come out of the fermentation lab. It's unique in that it's the only garum we produce that uses both natural enzymes from the animals' digestive tract and those produced by koji to break down the squid's proteins.

The in-depth instructions for Beef Garum (page 373) serve as a template for all the garum recipes in this chapter. We recommend you read that recipe before starting in on this one.

Using a meat grinder, food processor, or blender, grind the squid into a rough puree. A meat grinder is the best option for this, but if you don't have one, chop the squid into manageable pieces before pulsing it in a food processor or blender. Transfer the pureed squid to a 3-liter food-safe fermentation vessel.

Next, grind (or blend) the barley koji and add it to the vessel, along with the water and salt. Stir the ingredients with a clean spoon and scrape down the inner sides of the container with gloved hands or a rubber spatula. Lay a sheet of plastic wrap directly on the surface of the liquid and cover the vessel loosely with a lid, leaving at least a corner slightly ajar. Ferment the

Squid Garum, day 1

Day 7

Day 75

garum in a fermentation chamber at 60°C/140°F or in an electric rice cooker on "keep warm" for 8 to 10 weeks, stirring once a week.

By the time it's done fermenting, the flesh of the squid should be almost completely broken down. It will smell like a pleasantly sticky marriage of the earth and the sea, and it should have a salty, umami taste that grips your taste buds.

To finish, you can do one of two things: (1) Line a fine-mesh sieve with cheesecloth, pour in the garum, and set it over a bowl for 24 hours to catch the liquid; or (2) puree the garum into a thick paste. If you choose the former option, you'll end up with two separate products—liquid garum and leftover solids—that can be used interchangeably; the paste will be better suited to being brushed over things like blanched asparagus, whereas the liquid will easily dissolve into stocks or broths.

Pour the garum into bottles or another covered container. The garum is very stable and will keep well in the fridge for months. You can also freeze it for longer storage without any negative effects, but note that because of the high salt content, it probably won't freeze completely solid.

Suggested Uses

Pissaladière

Squid garum accentuates already fishy flavors in a really elegant way, and its saline funk is excellent at stringing together sugary sweetness with earthier tastes. In other words, it's an ideal addition to pissaladière, a favorite rustic dish of many Noma chefs who have traveled through France. Pissaladière is a *pâte brisée* baked like a pizza and coated with caramelized onions topped with black olives and anchovies. The straightforward Niçoise classic is helped greatly by adding a tablespoon of squid garum to the onions just as they've finished caramelizing in the pan.

Crudités

Squid garums can help transform raw vegetables into a complete dish. Dress a plate of crudités—Romanesco, carrots, and radishes, perhaps—with olive oil, sea salt, a pinch of red pepper flakes, and a very light drizzle of squid garum for an appetizer that is funky and refreshing at the same time.

A splash of squid garum enriches caramelized onions for a pissaladière.

387

Our goal for roasted chicken wing garum was to see
what would happen if we fermented a meat product
that was already at the height of its deliciousness.

388

Roasted Chicken Wing Garum

Makes about 1.5 liters

2 kilograms chicken bones
3 kilograms chicken wings
450 grams Pearl Barley Koji (page 231)
480 grams non-iodized salt

Roasting brings a lot of rich, fully developed flavor to this garum, meaning it needs only about a month of fermentation to coax out more umami. If we were to ferment this chicken garum as long as we do beef or squid garum, it would lose its subtlety and complexity.

The in-depth instructions for Beef Garum (page 373) serve as a template for all the garum recipes in this chapter. We recommend you read that recipe before starting in on this one.

Place the bones in a large pot and fill with water just to cover—about 3 liters. Bring the water to a boil, skimming away any impurities that float to the surface as it comes to temperature. Once it reaches a boil, reduce the heat to a simmer and cook the stock for 3 hours.

In the meantime, heat the oven to 180°C/355°F. Line a baking sheet with parchment paper. Place the chicken wings on the lined sheet and roast them for 40 to 50 minutes, tossing several times while cooking to ensure that they get an even, dark browning.

Remove the wings from the oven and let them cool down. Weigh out 2 kilograms of the roasted wings and use a cleaver to chop them into small pieces. (If you have any extra wings, well, then have yourself a snack.)

Roasted Chicken Wing Garum,
day 1

Day 7

Day 30

Strain the chicken stock through a fine-mesh sieve and allow it to cool.

Pulse the koji in a food processor to break it up into small pieces. Put the chopped chicken wings, koji, salt, and 1.6 kilograms of the chicken stock in a 3-liter fermentation vessel of your choice and stir to combine thoroughly. Scrape down the inner sides of the container with gloved hands or a rubber spatula and lay a sheet of plastic wrap directly on the surface of the liquid. Cover the container with a lid; screw it on slightly less than completely tight if it's a screw cap or leave it slightly ajar in one corner if it's a snap lid. Ferment the garum in a fermentation chamber at 60°C/140°F or in an electric rice cooker on "keep warm" for 4 weeks.

Every day for the first week, use a clean spoon or ladle to skim off as much fat as you can, then stir the garum and cover again. After the first week, skim and stir once a week.

To harvest, pass the garum through a fine-mesh sieve, and then again through a sieve lined with cheesecloth. Allow the liquid to settle and skim off any fat that floats to the surface.

Pour the garum into bottles or another covered container. The garum is very stable and will keep well in the fridge for months. You can also freeze it for longer storage without any negative effects, but note that because of the high salt content, it probably won't freeze completely solid.

Suggested Uses

Ramen Broth

When first tasting roasted chicken wing garum, almost every Noma chef mutters the same word: "Ramen." It's true, this garum possesses some of the same deep, meaty tones of a great bowl of ramen. A splash poured into a basic kombu and katsuobushi dashi makes for a convincing cheat. And if you've made a more proper ramen broth, a touch of garum will help kick the flavor up to eleven.

Roasted Cashews

Coat cashews (or any nut of your choice) with melted butter and spread onto a baking sheet or oven-safe pan. Roast in a 160°C/320°F oven until they become golden brown and fragrant. Remove them from the oven and mix in a couple of tablespoons of chicken wing garum. Don't add so much garum that the liquid pools on the pan. All the garum should be absorbed by the nuts and evaporated by the heat. You don't want the cashews to become soggy. Once they cool, they should still be crunchy, with a savory, salty crust.

Toss pan-roasted nuts in chicken wing garum for an indescribably delicious snack.

391

Grasshopper garum is a continuation of years of
cooking with insects at Noma.

Grasshopper Garum

Makes about 2 liters

600 grams grasshoppers or crickets
  (live or dead)
400 grams wax moth larvae
225 grams Pearl Barley Koji (page 231)
800 grams water
240 grams non-iodized salt

Grasshopper garum is by far the most magical ferment in this book, not least because it will instantaneously remove any mental block you have about cooking with or eating insects. We're almost reluctant to use it when we're developing recipes because it's so good, it almost feels like a crutch.

Grasshoppers can be purchased through pet stores or edible-insect companies. The wax moth larvae called for may be more difficult to procure, in which case you can omit them and replace their weight with more grasshoppers; the finished product will be slightly less rich but no less delicious. (If you can't find grasshoppers, crickets will work too.)

The in-depth instructions for Beef Garum (page 373) serve as a template for all the garum recipes in this chapter. We recommend you read that recipe before starting in on this one.

Blend the grasshoppers and larvae into a paste and transfer to a bowl. Pulse the koji in a food processor, to break it up into small pieces. Fold together the insects, koji, water, and salt, then transfer the mixture to a 3-liter fermentation vessel of your choice. Scrape down the inner sides of the container with gloved hands or a rubber spatula and lay a sheet of plastic wrap directly on the surface of the liquid. Cover the container with a lid; screw it on slightly less than completely tight if it's a screw cap or leave it slightly ajar in one corner if it's a snap lid.

393

Grasshopper Garum, day 1

Day 7

Ferment the garum in a fermentation chamber at 60°C/140°F or in an electric rice cooker on "keep warm" for 10 weeks, stirring once a week. It's finished once the mixture tastes nutty, toasty, and packed with umami.

To harvest, puree the garum into a fine paste and pass it through a fine-mesh sieve or tamis. Pour the garum into bottles or another covered container. The garum is very stable and will keep well in the fridge for a month. You can also freeze it for longer storage without any negative effects.

Suggested Uses

Grasshopper Butter

Bring a stick of butter to room temperature, then whisk it together with 20% grasshopper garum by weight. Transfer to an airtight container and store in the fridge. You can use grasshopper butter in any savory application that calls for regular butter: roasting vegetables, basting meats or fish, even cooking pancakes. Speaking of which . . .

Savory Pancakes

Omit the sugar from your favorite pancake recipe and cook them in grasshopper butter. Brush the finished pancakes with a bit of grasshopper garum, then fold them over and fill them with chopped red onions, a dollop of crème fraîche, and a spoonful of good-quality caviar or fish roe. (When it comes to fish eggs, it's all about the quality and not necessarily the type. Fresh, high-quality lumpfish, salmon, or trout roe is much preferable to mediocre caviar.) Finish with fresh chives. People will go nuts.

Day 75

Bee pollen is an incredibly complex ingredient,
with flavors that derive from the flowers near the
bees' hive.

396

Bee Pollen Garum

Makes about 1.5 liters

1 kilogram fresh or frozen bee pollen
200 grams Pearl Barley Koji (page 231)
300 to 600 grams water
60 grams non-iodized salt

Once again, we turn our attention to the underappreciated edible insect kingdom. Bee pollen is extremely complex, chemically speaking, housing dozens of species of fungi and bacteria. It's extremely sweet, and can sometimes comprise more than 50 percent protein. The composition and flavor of the pollen can vary wildly, though, depending on what flowers the bees have been harvesting from. It's also more readily available than you might think, as bee pollen is often sold as a nutritional supplement and can be ordered through health food suppliers (see Sources, page 448).

The in-depth instructions for Beef Garum (page 373) serve as a template for all the garum recipes in this chapter. We recommend you read that recipe before starting in on this one.

If the pollen you procured is dried, first puree 700 grams of it in a blender with 300 grams water to achieve the same moisture content as fresh bee pollen.

In a blender, combine the pollen (or pollen puree), koji, 300 grams water, and salt and blend until smooth, then transfer the mixture to a 3-liter fermentation vessel of your choice. Scrape down the inner sides of the container with gloved hands or a rubber spatula and lay a sheet of plastic wrap directly on the surface of the liquid. Cover the vessel tightly with a lid or more plastic wrap.

397

Bee Pollen Garum, day 1

Day 7

Day 21

Ferment the garum in a fermentation chamber at 60°C/140°F for 3 weeks, stirring once a week. The amount of sugar in the pollen means it will brown and caramelize much faster than other garums, and therefore doesn't require as much time in the heat.

To harvest, puree the garum into a fine paste and pass it through a fine-mesh sieve. Pour the garum into jars or another covered container. The garum is very stable and will keep well in the fridge for a month. You can also freeze it for longer storage without any negative effects.

Suggested Uses

Pollen Oil

When we were first trying to understand the applications of bee pollen garum in the fermentation lab, a little research taught us about bee pollen's solubility in fat. That nugget of knowledge led to a rather delicious experiment. Place 250 grams bee pollen garum in a blender with 500 grams neutral vegetable oil—rapeseed and grapeseed oil both work well. Blend the two together for 6 minutes, then transfer to a container and refrigerate overnight to infuse. The following day, after the heavier sediment has settled at the bottom of the container, gently pour the oil through a sieve lined with cheesecloth. The sediment that remains at the bottom of the container is still extremely delicious and should be saved for other uses. The floral, deep mustard-colored oil is subtly powerful, only becoming more flavorful the longer it rests on your tongue. It's an excellent replacement for the olive oil you'd use to dress beef tartare, or you can emulsify it with egg yolks for a very different but outstanding mayonnaise. Or simply toss roasted root vegetables, like celery root or sweet potatoes, in a bit of bee pollen oil as they come out of the oven.

Bee Pollen Risotto

Bee pollen garum is probably the only one we make at Noma that is mild enough to eat by the spoonful. A while back, when Michelin three-star chef Massimiliano Alajmo of Le Calandre in Sarmeola di Rubano, Italy, was visiting the restaurant as a guest chef, he made a risotto with bee pollen garum in place of cheese that our guests fell completely in love with. You can do the same at home by cooking a classic risotto with onions, white wine, and chicken stock, then finishing it with about 2 tablespoons of bee pollen garum per portion. The garum takes the dish in a wildly different but somehow familiar direction.

Roasted Tomatoes

In the late summer, when the tomatoes can't get any better, take a basket of floral and fragrant cherry tomatoes—Sun Golds are a really spectacular cultivar—and toss them into a scorching-hot pan coated with olive oil. Toss them around for a few seconds before sliding the pan onto the top rack of the oven to finish cooking beneath the hot broiler. Let the tomatoes bubble and burst and caramelize until you see their juices thickening at the bottom of the pan, about 10 minutes. Pull the pan from the oven, throw in a couple of sprigs of lemon thyme, and add a healthy spoonful of bee pollen garum. Stir the mixture around quickly, then remove the lemon thyme sprigs and finish it off with torn leaves of opal basil. Serve it with crusty grilled bread, or strewn through dressed salad greens with sautéed chanterelles for a smashing warm summer salad.

399

Yeast Garum

Makes about 2.5 liters

300 grams fresh baker's yeast
725 grams nutritional yeast
250 grams Yellow Peaso (page 289)
225 grams Pearl Barley Koji (page 231)
1 kilogram water
200 grams non-iodized salt

All the other garums in this chapter have been made with animal proteins, but animals are not the only protein-rich organisms. Yeast, which we mostly use as an agent of fermentation, can also be fermented itself into a delicious vegan garum. Fresh baker's yeast can be found in the refrigerated section of good grocery stores or online, and is also sometimes sold as "compressed yeast" or "cake yeast." And if you haven't made the Yellow Peaso this recipe calls for, you can substitute store-bought miso; aim for a lighter variety like okasan or shiro miso.

The in-depth instructions for Beef Garum (page 373) serve as a template for all the garum recipes in this chapter. We recommend you read that recipe before starting in on this one.

Heat the oven to 160°C/320°F. Line a baking sheet with parchment paper. Crumble the baker's yeast onto the lined baking sheet and roast it until it turns deep brown and smells nutty and meaty, about 1 hour. Remove the yeast from the oven and allow it to cool.

The baker's yeast will lose a lot of weight in the oven; measure 75 grams of it and combine with the rest of the ingredients in a blender. Blend all the ingredients together until you have a smooth paste, about 45 seconds. Transfer to a 3-liter fermentation vessel of your choice. Scrape down the inner sides of the container with gloved hands or a rubber spatula and lay a sheet of plastic wrap directly on the surface of the liquid.

Yeast Garum, day 1

Day 7

Day 30

Cover the vessel tightly with a lid or more plastic wrap. Ferment the garum in a fermentation chamber at 60°C/140°F or in an electric rice cooker on "keep warm" for 4 weeks, stirring once a week. Once finished, the garum should be meaty, rich, sour, and full of umami, with a gorgeous, shiny, deep brown hue.

To harvest, puree the garum into a fine paste and pass it through a fine-mesh sieve lined with cheesecloth. Spoon the thick garum into jars or another covered container. The garum is very stable and will keep well in the fridge for months. You can also freeze it for longer storage without any negative effects.

Suggested Use

Smoked Hummus

While Noma's second iteration was under construction on the outskirts of Copenhagen's Christiania neighborhood, we held a series of casual pop-up dinners by the canal in the center of the city. Our Lebanese chef de partie, Tarek Alameddine, made this incredible hummus for guests to snack on while sitting at the wine bar. First, cold-smoke the chickpeas with hay (standard wood chips will work, too) for about 1 hour. From there, it's a pretty straightforward hummus recipe: Blend 500 grams smoked chickpeas with 75 grams tahini, 75 grams yeast garum, a clove of garlic, and a healthy splash of olive oil. Allow the machine to run for a good 5 minutes in order to yield a really smooth hummus. Finish it off with the zest and juice of 1 lemon.

Black Fruits and Vegetables

—

Black Garlic 417

Black Apples 425

Black Chestnuts 429

Black Hazelnuts 433

Waxed Black Shallots 437

404

Really Slow Cooking

Black fruits and vegetables have been part of the Western cooking repertoire for only about two decades now. At Noma, we got our first taste of black garlic—the prototypical blackened vegetable—about fifteen years ago, but we've only recently begun to experiment with the process of blackening other fruits, vegetables, and nuts.

To be clear, blackening is not fermentation. It is in large part an enzymatic process, and while all fermentation processes are enzymatic, not all enzymatic processes are fermentation processes. Still, we believe that blackening has a place in this book because it shares the same transformative magic as microbial fermentation, and because the products it creates have a similar place in our pantry.

The actual process of blackening is a sight to behold. Vegetables slowly morph into ripe fruits, as sharp flavors mellow and hard textures end up putty-like and malleable. Take black garlic, for instance. As with so many ferments, there's a wide range of quality in black garlics—some varieties can taste too raw and are not particularly enjoyable—but at its best, it's like the perfect grown-up candy, sweet and leathery and full of complexity.

To make black garlic, all you have to do is place a head of garlic in a sealed container and keep it at a constant temperature of 60°C/140°F for 6 to 8 weeks. That's the whole procedure—at least at the broadest level.

Temperature is the reason why blackening is not a fermentative process. The fungi and bacteria that we harness for fermentation can't survive at 60°C/140°F, so in the absence of microbial activity, we're left with only chemical processes.

Perhaps the easiest way to describe blackening is as very slow, very dark browning. Many readers of this book may be familiar with the Maillard reaction—the phenomenon responsible for crusty steaks, browned onions, toast, coffee, and innumerable other pillars of cuisine. The Maillard reaction is one of several forms of browning at work in black fruits and vegetables.

Soft-Boiled Egg and Black Garlic, Noma, 2012

A soft-boiled egg is dressed with nasturtium leaves and served in a bowl painted with a paste of black garlic blended with fermented honey, peas, and koji.

405

There's also caramelization, which is the pyrolysis of sugar. As heat is applied to organic compounds in the absence of oxygen, they thermally decompose. That's pyrolysis. Caramelization produces a flight of volatile flavors and aromas, as well as the lovely range of colors we tend to associate with delicious things.

We're accustomed to the Maillard reaction, pyrolysis, and caramelization taking place over a short time at high heat—usually 170°C/340°F or higher. But high heat is not explicitly necessary for any of these reactions to take place, if you're patient. The point of blackening is to stretch out the process over weeks. This is possible because the temperature of an object is actually the average of *billions of billions* of molecules moving around at different speeds. In a clove of garlic held at 60°C/140°F, 99.9999 percent of the molecules might very well be moving too slowly to inspire a pyrolytic or Maillard reaction. But occasionally, one molecule out of a billion might be moving fast enough to spark one of these energetic chemical reactions. From there, these isolated, sparse reactions cascade.

As pyrolysis breaks down larger sugars into smaller parts, it liberates more molecules, which are then available to take part in further reactions. Over the course of weeks, the products of these rare and irreversible chemical reactions accumulate, producing the dark sweetness we associate with black garlic. In fact, left long at this temperature, the garlic will eventually burn.

Pyrolysis, caramelization, and the Maillard reaction are non-enzymatic browning reactions. Enzymatic browning, on the other hand, is what happens to fruits and vegetables as they ripen and age over time. The enzyme polyphenol oxidase is important to a plant's health as it grows, but once a fruit's or vegetable's flesh is exposed to oxygen, the enzyme begins altering phenolic compounds in the plant's tissue and producing melanins, which are responsible for fruit turning brown. (Phenols are a large group of compounds, some of which give the flesh and fruit skins of many plants their color, and some play a big role in their flavor and aroma as well.) In a healthy plant, the creation of melanins at the site of a bruise or trauma helps to ward off infection, as melanins have antibacterial

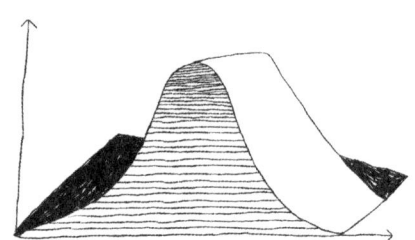

Temperature is an average of the kinetic energy of countless molecules moving at different speeds.

properties. In the blackening process, they serve to further darken the fruit or vegetable.

When you blacken fruit, both enzymatic and non-enzymatic reactions are taking place. It's unclear who first fused these various processes to blacken food, though most signs point to the storied fermenting culture of Korea, where centuries ago people were aging whole heads of garlic in earthenware pots during the hot summer months. As for the modern history of black fruits and vegetables, we again look to Korea, in the year 2004. The contemporary popularization of black garlic is credited to a man named Scott Kim, who devised a simple method for making black garlic using heat- and humidity-controlled aging chambers. We use the exact same methods at Noma to maintain an environment that leads to a cascade of slow chemical reactions, completely transforming ordinary ingredients over the course of weeks or months.

Redox Redux

In the vinegar chapter (page 157), we met Antoine Lavoisier, the father of modern chemistry and the man responsible for identifying oxygen as the reagent responsible for combustion. However, he wasn't actually the first person to publish research on the isolation of oxygen molecules. The English chemist Joseph Priestly, a collaborator of Lavoisier's, actually takes the credit for that. Priestly heated a metallic compound—mercury oxide, then known as "red calx"—and observed that as the air in the flask was becoming more flammable, the calx itself was reducing in weight. What Priestly witnessed was a *reducing* reaction (named because of the calx's reduction in weight). Once heated, the mercury oxide (HgO) separated into its two constituent elements, releasing its oxygen and rendering the air more combustible.

What's happening on the molecular level is an exchange: When mercury and oxygen first join to form mercury oxide, two atoms of mercury are surrendering an electron each to the oxygen. When the compound decomposes, the process is reversed. The electrons go back to their original owners, and gaseous oxygen is released into the air. Scientists initially believed that such reactions dealt exclusively with oxygen,

407

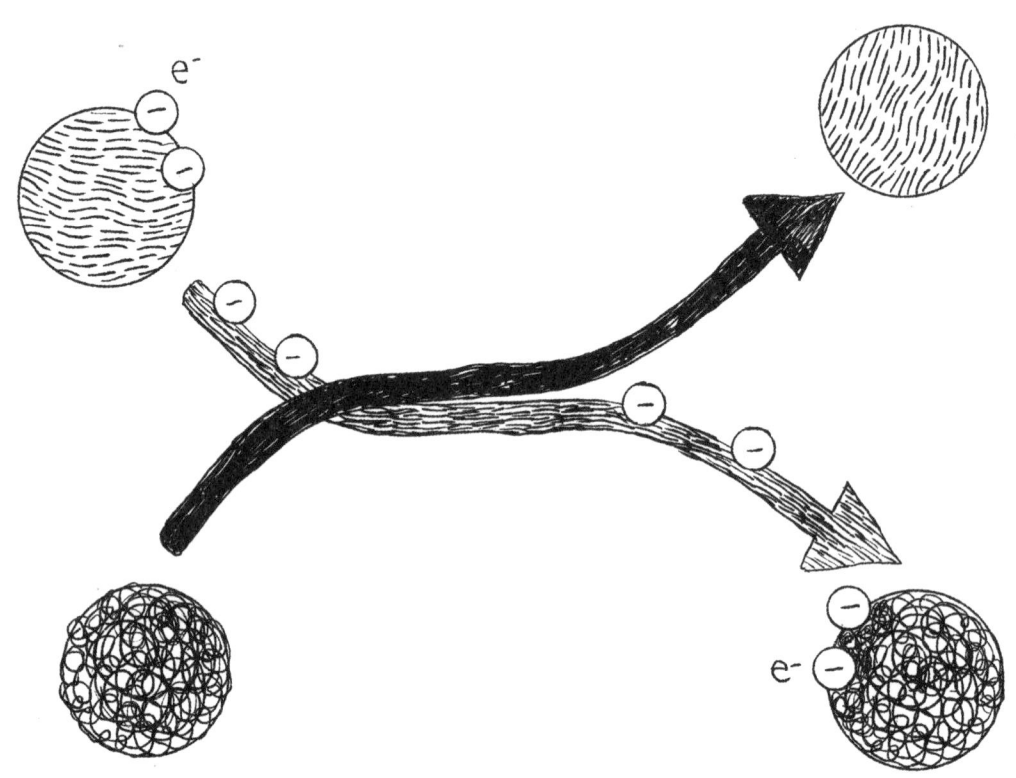

408

The French physician Louis Camille Maillard first unlocked the secrets of browning. Maillard reactions are redox reactions that can take place between reducing sugars (found in starches) as well as amino acids (found in proteins).

and so, to this day, the act of a molecule or atom being stripped of an electron is known as "oxidation." In Priestly's experiment, the mercury is being *oxidized*, and the mercury is *reducing* the oxygen. Parallel chemical reactions like this are known as *redox* reactions. Many of them involve oxygen, but there are plenty of other elements and compounds that can take part in redox reactions. However, one thing to note about redox reactions is that they are a two-way street. There is no oxidation without simultaneous reduction, and vice versa. Any time you hear about something being oxidized, you can be sure that something else is being reduced.

The Maillard reaction is a genre of redox reaction that is most frequently observed when cooking foods at high temperatures. It owes its name to Louis Camille Maillard, who was a young French physician when he discovered the process in the early 1900s at the University of Paris.

As mentioned above, there are many other elements and compounds besides oxygen that can participate in a redox reaction. This includes amino acids. When food is heated, sugars like fructose or glucose or those bound up in starch take part in redox reactions with free amino acids or those bound in protein chains. The reactions lead to highly unstable intermediary products that then further break down in a variety of ways, creating flavor compounds responsible for the color and delicious taste of browned food. The crust of bread, the seared surface of a scallop, brown butter—all these are products of the Maillard reaction. (It won't necessarily come in handy here, but the reaction is expedited by alkaline environments, as observed in the crust of pretzels, which are soaked in a dilute lye solution before baking.) Different flavors will develop depending on which amino acids are present. The artificial-flavor industry accomplishes a lot of its work by choosing specific amino acids to undergo redox reactions.

Most Maillard reactions take place at temperatures well in excess of 115°C/239°F, when there is enough kinetic energy to force the reagents to interact. Water is so good at absorbing heat energy that regardless of how much thermal energy

409

410

Black pears are blended into a paste and
dried into fruit leather before being
molded into imitation mussel shells and
filled with kelp ice cream and licorice.

Running Your Own Blackening Experiments

you add, mixtures containing water don't tend to rise above water's boiling point of 100°C/212°F until most of the water has boiled away. Therefore, in watery environments, it takes much longer for a Maillard reaction to take place. (Beyond this thermal effect, water can also inhibit the formation of certain products of the Maillard reaction.) But as mentioned earlier in this chapter and in the miso chapter (page 269), given enough time, the Maillard reaction will occur, even at low temperatures and in wet environments.

That might be a lot of chemistry to take in, but cooking *is* chemistry. Every time you light your barbecue, bake a cake, or cure a ham, you're carrying out multiple chemical reactions. Browning (and blackening) food is certainly all about chemistry. The Maillard reaction is only one of the chemical processes at work in a blackening chamber. As always, it's worth having a grasp on the science to help you make adjustments or improvements down the road.

The truth is, we're still learning how to put blackened fruits and vegetables to full use at Noma. We've only scratched the surface of all the different products that can be made by blackening. But the successes we've had so far have been encouraging. Blackened apples are amazing, for instance, especially after you dry them. The juice that leaches out of the apples as they blacken is beautiful as well—you can drink it all on its own, or ferment it into black cider. At the restaurant, we've served an invigorating first course of fresh apples marinated in a paste of black apples from the previous year that really highlights the contrast between blackened and fresh fruit.

It's not clear what other ingredients might benefit from blackening, but such is the beauty of exploration. It's a field that has not been studied very much at all. The person who has pushed us the most in this realm is Matt Orlando of Amass restaurant in Copenhagen. He's experimented a lot with blackening, and it's been very inspiring to follow him—he's way ahead of the curve.

411

Of course, there's an entire planet of ingredients that remain unblackened, waiting for you to give them a whirl. In our fermentation lab, we've put a lot of different fruits and vegetables through this process, from cabbage (not so good) to corn (so-so) to chestnuts (so, so good!). Trial and error has edged us toward a set of parameters that seem to dictate whether a particular fruit or vegetable makes for a worthy blackening candidate. Garlic, the quintessential blackened vegetable, actually sheds a lot of light on this subject. Each of the following factors explains why garlic is particularly well suited for blackening, and what you should look for in other things you want to blacken.

Moisture Content and Retention

Moisture is important to the enzymatic browning of vegetables and fruits—that's why dehydration is such an effective means of preservation. If an ingredient is too dry, it won't blacken. Garlic isn't an especially wet ingredient, but it's good at retaining what moisture it has. The multiple layers of skin make for a strong guard against rapid moisture loss, and while the garlic never dries out completely, it dries enough to facilitate the Maillard reaction.

Fruits or vegetables that contain large amounts of water will blacken, but you'll face a different challenge: For the most part, the thin skins of fruits like apples and pears will give out and their flesh will disintegrate. They lose their shape long before any interesting flavors show up. This problem can be mitigated, as you'll see in Black Apples (page 425), but something like garlic naturally has the right water content and defense against moisture loss to deliver a final product that has tons of character and holds on to its form.

Sugar Content

If you've ever crushed a clove of garlic with the broad side of a knife, you're familiar with the sticky, tacky paste you get as a result. This happens because garlic, like many other members of the allium family, contains great stores of sugar. Because of its pungency, you might not think of raw garlic as sweet, but its plentiful sugars are necessary for the slow Maillard reaction and caramelization that take place during blackening. Ingredients lacking in sugar—take kohlrabi, for instance, something we thought would be promising as a blackened vegetable—taste pretty horrible when heat-treated. Without the harmonious flavor backbone created by the Maillard reaction, blackening can create heavy, acrid notes from pyrolysis, with no sweetness to temper them.

Pungency

Nothing comes out of the blackening chamber tasting the same. Sometimes that's a good thing, but other times the prolonged heat can mute or destroy subtle and volatile aromas. Raw garlic's pungency transforms during its two-month-long heat bath, but it stays recognizable. The subtlety of, say, a root of salsify is completely lost once it's blackened. Though the root is moist enough, has thick enough skin, and is also sweet enough, black salsify comes out delicious but indistinct. It ends up tasting more ambiguous than other roots. The more potent the flavor of the raw ingredient, the more interesting the outcome will be.

413

Sweet, pungent black garlic has grown in
popularity and prevalence in Western kitchens over
the past few decades.

Black
Garlic

Makes 10 heads

10 heads very fresh garlic

Black garlic is one of the easiest recipes in this book. The process runs itself, and while it's good to check on the progress as you go, there isn't much, if any, tending to be done. Your biggest challenge will be finding or creating a fermentation chamber that can hold a steady temperature of 60°C/140°F for weeks—that's the only way to spur all the redox, Maillard, enzymatic browning, pyrolytic, and other reactions you need to turn fruit and vegetables deliciously black.

Equipment Notes

Slow, consistent heat is the key to blackening garlic. In a restaurant, a warming cabinet is great for this. A homemade fermentation chamber (see "Building a Fermentation Chamber," page 42, as well as the instructions in the koji chapter, page 211) will work well, too. The simplest solution, however, is an electric rice cooker or a slow cooker. The "keep warm" setting hovers right in the neighborhood of 60°C/140°F. They're not 100 percent accurate, but they should be adequate for small-batch blackening. However, not all electric cookers will stay on indefinitely. Make sure your cooker doesn't have an auto-off feature before jumping into the recipe.

417

Black Garlic, day 1

Day 7

In-Depth Instructions

Set up your fermentation chamber, which will need to maintain a steady temperature of 60°C/140°F for several weeks. If using an electric cooker, be aware that the heating elements in many cookers come into direct contact with the metal cooking bowl, which can cause the garlic to scorch, so place a small wire rack, plate, or bamboo mat in the bowl to act as a buffer.

Use freshly harvested summer garlic, which will have a good amount of moisture for the blackening process and won't have a sulfurous flavor. Stay away from garlic that has begun to sprout and display green shoots, but also be sure the cloves have fully developed, because the skin on very new garlic can be too papery. Avoid industrially produced white garlic from China, which can be acrid and sulfurous with very little sweetness. Elephant garlic is too mild to do well with this process either.

Peel off any dusty garlic skins and inspect the heads to make sure there is no mold hiding between the layers; if there is, peel away more skin and wipe it off. You'll now need to wrap the garlic to keep its moisture in. Industrial manufacturers use hermetically sealed rooms where the humidity is regulated. On a small scale, simply double-wrapping the heads in two large sheets of foil will be fine—be sure the heads sit in a single layer. (If you're blackening your garlic in a fermentation chamber, like the ones suggested for koji but with the heat cranked way up, place the foil-wrapped heads of garlic in a larger airtight container, such as a large zip-top bag or a plastic snap-lid container.) Alternatively, you could seal the heads in a vacuum bag with the sealer at 50% suction (lest the bag get punctured by the garlic stems) or use two heavy-duty zip-top plastic bags with as much air squeezed out as possible.

Place the wrapped garlic in the fermentation chamber and close it, or place in the cooker and seal the lid. If the lid has a rubber seal with a latch, that's fantastic, as it will retain moisture well. If not, do your best to seal it some other way. While not exactly pretty, wrapping the top of the cooker in plastic wrap does the trick. Turn the cooker on to "keep warm," and away you go.

Check the garlic after a week to assess its progress. The outside of the skins should be starting to turn tan and look damp as they absorb moisture from the cloves within. If all's well and the garlic hasn't become overly dry, continue. If they've started to scorch or burn on the bottom, or if they've become extremely dry, unfortunately, you'll need to start again.

The total amount of time will range from 6 to 8 weeks for full development of flavor. When they're ready, the cloves should be black, slightly shrunken away from their skins, tacky to the touch, and easily squished between your thumb and forefinger. The flavor should be sweet, earthy, slightly fruity, and reminiscent of roasted garlic.

Let the whole heads rest on the counter at room temperature for a day to shed some of the residual moisture, then bag them or pack them in a covered container and store them in the fridge or freezer. Black garlic has less moisture and a pH well below that of fresh garlic, but it's not perfectly shelf stable, so it needs to be refrigerated or frozen. It will keep well for a week in the fridge; otherwise, freeze it for longer storage.

Day 30

Day 60

419

1. Vacuum-seal the garlic in vacuum bags.

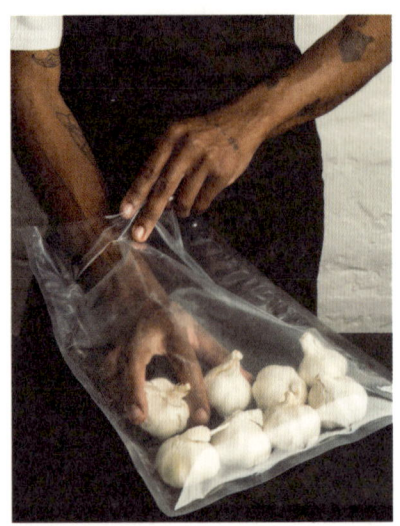

2. Alternatively, wrap the garlic in two layers of foil.

3. Garlic ready for blackening.

4. Place the garlic into your rice cooker or fermentation chamber.

5. Close the chamber and set it to 60°C/140°F, or seal the cooker and set to "keep warm."

6. The garlic will take 6 to 8 weeks to blacken.

Suggested Uses

Black Garlic Ice Cream

Black garlic can be employed in any part of a menu, from the very beginning to dessert. Buy or make a good-quality vanilla or chocolate ice cream, and fold a spoonful of chopped black garlic into each serving. Chocolate ice cream is further improved by the smallest drizzle of olive oil, too.

Black Garlic Skin Broth

When you've harvested all the flesh from your black garlic, don't throw away the skins. Save them to add by the fistful to chicken broth—or any broth, for that matter—during the last hour of cooking for extra richness, depth of flavor, and a delicious fruitiness. In fact, if you're planning to add black garlic skins to your broth, don't even bother using other vegetables at the beginning. The skins will have more than enough flavor to complement the chicken on their own. We love to do this at Noma for staff meal, especially at breakfast with a few drops of chili oil and maybe some rice or noodles on the side.

Black Garlic Paste

Crush peeled black garlic with a mortar and pestle and a bit of water or oil until you have a thick paste with plenty of possible uses. For starters, it makes a fine accompaniment to sliced hard cheeses, just like quince paste. (If you find the black garlic paste a bit lacking in sweetness, add a drop of honey.) Should you mix equal quantities of black garlic and olive tapenade, you'll wind up with a delicious topping for a slab of crunchy-soft bread that's been toasted with plenty of olive oil. Last but not least, when black garlic finds its way into pesto (about a teaspoon per serving) and that pesto finds its way into your pasta, you will not be disappointed.

Crush black garlic with a mortar and pestle,
then stir it into ice cream for an unexpectedly
well-matched combination.

423

The tartness of a good apple sits in pleasant contrast
to the deep sweetness produced by blackening.

424

Black
Apples

Makes 10 apples

10 apples

Apples tend to fall apart when exposed to moist heat, which poses a problem for the blackening process we use for garlic. As a workaround, we carefully blacken the apples, then remove the moisture afterward with a dehydrator. This technique works equally well with pears, quince, and other pome fruits. The blackened results are deeply flavorful, sweet, and chewy like toffee.

The in-depth instructions for Black Garlic (page 417) serve as a template for all the blackened fruit and vegetable recipes in this chapter. While the method differs somewhat here, we recommend you read that recipe before starting in on this one.

Peel the apples and arrange them in a large vacuum bag so they're not touching. The apples will need to sit flat in your fermentation chamber, so if it looks like your chamber might be too small, remove a few apples. Seal the bag on maximum suction. You can also use a large zip-top bag and squeeze all the air out by placing the apples into the bag, then slowly lowering it into a tub of water, stopping a few centimeters from the top (you may need to pull from the bottom of the bag to counteract the fruit's buoyancy). The pressure of the water will force the air out. Seal it shut and you'll have an effective, albeit imperfect, vacuum.

Black Apples, day 1

Day 7

Day 60

Place the apples in the fermentation chamber. If you're using an electric rice cooker or slow cooker, remember to raise the fruit off the bottom with a plate, wire rack, or bamboo mat. Close the chamber and set it to 60°C/140°F, or seal the cooker and turn it to the "keep warm" setting.

Within the first few days, the apples will turn light brown and begin to leach their juices, which will collect at the bottom of the bag. The best possible thing you can do? Don't disturb them! After the first couple of weeks, the flesh will have degraded so much that even a little jostling may wreck the apples' structure. Leave them alone for a total of 8 weeks, by which point they will have blackened quite a bit.

The apples will be extremely fragile at this point, so carefully remove the bag from the fermentation chamber and cut it open, draining and reserving the liquid. Using a spoon or spatula, transfer the apples to a dehydrator tray or baking sheet lined with parchment paper. Dehydrate the apples at 40°C/104°F or in a very low oven; they should take 24 to 36 hours to dry to the proper doneness. Rotate the apples a couple of times throughout the process so they dry evenly. They're done when their texture is chewy like toffee. Once dried, they can be stored covered in the fridge for a week, or frozen for longer storage.

Dried black apples dipped in chocolate.

Black apples impart their flavor to brandy (while absorbing some of the brandy's flavor themselves).

Suggested Uses

Black Apple Leather

Like dried apricots, dried black apples take on a texture similar to a chewy toffee. Once dried, the apples make a delightful snack on their own. You could dry the apples whole for a couple of days in a dehydrator, but if you want things to move more quickly and you don't mind losing the apple shape, puree the fruit without its juice, then spread the puree into a layer 3 millimeters / ⅛ inch thick on a nonstick mat before drying into sheets. Chewy black apples (or black apple leather) are primed to dress in the Mexican way of eating fruit, by squeezing a little bit of lime or lemon juice over the top and sprinkling with chile salt.

Chocolate-Covered Black Apples

Begin by slicing the cheeks of the dried apple away from the core (if you used a smaller variety of apple, leave them whole). Temper a good-quality dark chocolate—at least 70% cacao—then dip the apple slices into the chocolate. Allow to cool on a rack or parchment paper so that the chocolate coating becomes a crunchy contrast to the rich, creamy texture of the apple inside.

Brandied Black Apples

This makes for a fun gift, provided you plan well in advance. Place whole or halved black apples in a glass jar and cover with good brandy or Calvados. Seal and store in a cool place—the longer, the better. At the restaurant, we've sometimes let fruit sit in a boozy bath for up to two years. Once the alcohol has turned sweet and syrupy and the fruit is soft and alcoholic, it's time to break out the vanilla ice cream.

427

Find fresh chestnuts in the early
autumn, then transform them
through blackening.

428

Black
Chestnuts

Makes 1 kilogram

1 kilogram fresh chestnuts
   in the shell

Fresh, sweet chestnuts are at their peak in the early fall. They contain a fair amount of water, and although they have a shell to keep that moisture in, like garlic, they should be wrapped in foil or plastic for more moisture retention. At Noma, we've found that chestnuts taste far more interesting when they're not fully blackened. Held at 60°C/140°F, the chestnuts take about 4 weeks to mature to the ideal point. The flavor is akin to grape must, with notes of plums and dried fruit. Any chalkiness you might associate with raw chestnuts gives way to a pleasantly meaty texture with a bit of snap. Past the fourth week, deep caramel tones set in but the flavor becomes a bit one-dimensional.

The in-depth instructions for Black Garlic (page 417) serve as a template for all the blackened fruit and vegetable recipes in this chapter. We recommend you read that recipe before starting in on this one.

Arrange the chestnuts in a single layer in a vacuum bag. They'll need to sit flat in your fermentation chamber, so if it looks like your chamber might be too small, remove a few chestnuts. Seal the bag on maximum suction. You can also use a large zip-top bag and squeeze all the air out by placing the chestnuts in the bag, then slowly lowering it into a large tub of water, stopping a few centimeters from the top (you may need to pull from the bottom of the bag to counteract the chestnuts' buoyancy). The pressure of the water will force the air out. Seal it shut and you'll have an effective, albeit imperfect, vacuum.

Black Chestnuts, day 1

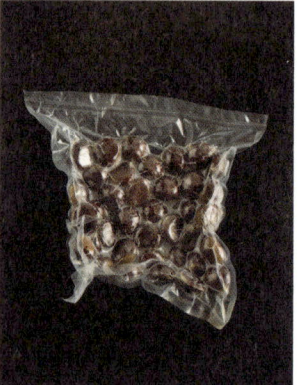

Day 14

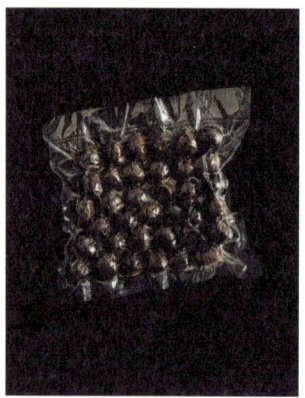

Day 30

Place the chestnuts in the fermentation chamber. If you're using an electric rice cooker or slow cooker, remember to raise the chestnuts off the bottom with a plate, wire rack, or bamboo mat. Close the chamber and set it to 60°C/140°F, or seal the cooker and turn it to the "keep warm" setting.

Leave the chestnuts in the chamber or cooker for 4 weeks. Slit one open to taste and decide whether you want to go a little longer. Once they've blackened to your liking, keep them in the shell until you want to use them. Store in a sealed container in the fridge if you plan to use them within a week, or freeze for longer storage.

Suggested Use

Stuffed Pasta

With very little modification, black chestnuts can be turned into an unbelievable filling for pasta. Start by chopping 350 grams peeled black chestnuts into thin slivers. In a medium sauté pan, heat 100 grams butter until it begins to foam. Throw the chestnuts in and sauté for a few minutes before adding 250 grams good chicken stock. Cook at a simmer beneath a round of parchment paper until the chestnuts are fairly tender. At that point, transfer the contents of the pot to a blender and puree the mixture until it's silky smooth (depending on your blender, it may need a touch of water to help it spin). Season the puree with salt and mace or nutmeg. Pipe it onto fresh pasta and form your preferred stuffed shape: cappelletti, agnolotti, tortellini, ravioli. And to take things to an entirely new dimension, glaze your pasta with the Lacto Koji Butter Sauce (page 261) after it's cooked.

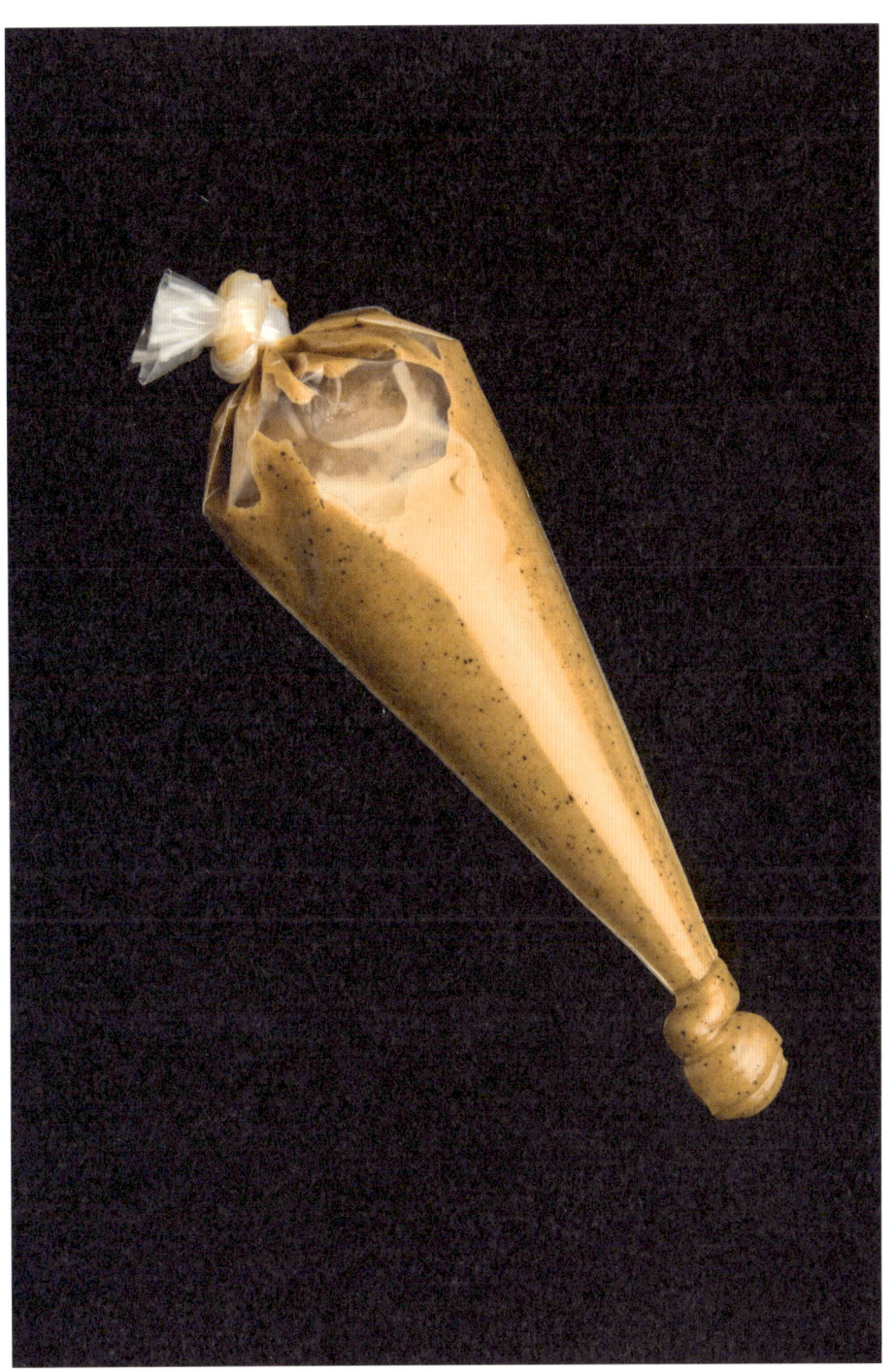

Puree black chestnuts with a bit of chicken stock
for a rich and nuanced pasta filling.

431

Fresh hazelnuts take on a toasty, chocolaty flavor
when exposed to prolonged heat.

Black Hazelnuts

Makes 1 kilogram

1 kilogram fresh hazelnuts in
the shell

Fresh hazelnuts (also called cobnuts or filberts, depending on the cultivar or region) can be ready to harvest as early as July, but definitely by August and early September. But beware: A hazelnut in its shell is not necessarily a fresh hazelnut—in fact, it probably isn't. In the Northern Hemisphere, hazelnuts are harvested once they've fallen from the tree and are often cellared for a while before reaching the consumer. This dries the nuts out completely, leaving inadequate moisture to facilitate blackening. When picked straight from the tree, the flesh of hazelnuts is white and tender, with a snappy bite. The freshness and ripeness of your hazelnuts—they should be mature, not too young—is the biggest factor in making good black hazelnuts, so you'll probably need to find an obliging grower or hope you get lucky at the farmers' market.

The in-depth instructions for Black Garlic (page 417) serve as a template for all the blackened fruit and vegetable recipes in this chapter. We recommend you read that recipe before starting in on this one.

Arrange the hazelnuts in a single layer in a vacuum bag. They'll need to sit flat in your fermentation chamber, so if it looks like your chamber might be too small, remove a few hazelnuts. Seal the bag on maximum suction. You can also use a large zip-top bag and squeeze all the air out by placing the hazelnuts in the bag, then slowly lowering it into a large tub of water, stopping a few centimeters from the top (you may need to pull

433

Black Hazelnuts, day 1

Day 7

Day 30

from the bottom of the bag to counteract the hazelnuts' buoyancy). The pressure of the water will force the air out. Seal it shut and you'll have an effective, albeit imperfect, vacuum.

Either way, the sharp tips of the hazelnuts can potentially puncture the bag once placed in the fermentation chamber, so it's a good idea to double-bag them.

Place the hazelnuts in the fermentation chamber. If you're using an electric rice cooker or slow cooker, remember to raise the chestnuts off the bottom with a plate, wire rack, or bamboo mat. Close the chamber and set it to 60°C/140°F, or seal the cooker and turn it to the "keep warm" setting.

Leave the hazelnuts in the chamber or cooker for 4 to 6 weeks, during which time their flesh will shrink a little. Crack one open and have a look. It should have a deep golden to dark brown color. The taste is something remarkable, like a cup of hot chocolate with a spoonful of Nutella in it. The texture will no longer snap as it did when fresh, but will have a pleasant chewiness. Use immediately or store them, sealed, in the freezer so they don't dry out, which affects their taste dramatically.

Suggested Uses

Sole à la Grenobloise

Black hazelnuts taste remarkably like hot chocolate, and their sweet, deeply roasted flavor makes them extremely fun to cook with. They're a perfect unexpected companion to dishes like *sole à la grenobloise*. Start by cracking black hazelnuts out of their shells and chopping them into a coarse meal—you'll need about 30 grams. Dredge skinless sole fillets in regular flour, then fry them in a hot pan with a good amount of olive oil until golden brown, about 90 seconds on each side. While the fish is cooking, put 30 grams butter in a smaller saucepan and let it sizzle and brown, then throw in the black hazelnuts and stir until they become aromatic. Add 30 grams chopped capers,

10 grams chopped parsley, and the juice of 1 lemon, and swirl the pan to form a quick sauce. Drown the browned fillets of sole in the nutty Grenobloise and serve immediately.

Black Hazelnut Milk

Nut milk is a beverage that's growing in popularity, and black hazelnuts make an especially amazing nut milk. Crack a handful of nuts out of their shells and cover with twice their weight in water. Allow the nuts to soak in the fridge overnight. The following day, blend them in a blender until completely smooth, about 3 minutes. Strain the mixture through a sieve or chinois lined with a couple of layers of cheesecloth, pressing down on the pulp with a ladle to extract as much milk as possible. From there, use your black hazelnut milk to make hot chocolate or horchata (replacing the water or milk with hazelnut milk), or garnish it with a splash of hazelnut oil and use it as a sauce for seared scallops.

Puree the nuts and water, then strain through a sieve to harvest the milk.

Coating shallots in wax before blackening seals in
moisture and imparts a pleasant honey flavor.

436

Waxed Black Shallots

Makes 1 kilogram

500 grams beeswax
1 kilogram fresh shallots

We've seen how to blacken ingredients in impermeable bags and foil, but these aren't the only methods. Plastic is useful for small products like nuts, but for larger items like shallots, encasing them in wax is an interesting alternative that imparts its own flavor to the final blackened product. Look for an organic food-grade beeswax (available online).

The slim margin between the ideal blackening temperature (60°C/140°F) and the temperature at which the beeswax (which covers the shallots) will melt (64°C/147°F) can make the difference between beautifully blackened shallots and a terrible mess. A rice cooker set to "keep warm" may not be precise enough to pull this off. Therefore, you'll need a fermentation chamber with more precise temperature control. (See "Building a Fermentation Chamber," page 42.)

In a very small pot over medium heat, melt the beeswax. The wax should only *just* be melted; don't heat it up more than needed to keep it fluid. You'll want the wax as deep as possible in order to fully immerse the shallots, so the smaller the pot's diameter and the taller its walls, the better.

Peel the shallots, keeping the root intact. Make sure the shallots are moist and don't have any visible mold on them. Working one at a time, pick up a shallot by piercing it in its root end with a skewer or tweezers. Dip the shallot into the wax briefly and pull it out, letting the excess drip back into the pot.

437

Black Shallots, day 1

Day 7

Day 60

Hold it in the open air until the wax turns opaque and solidifies; it won't take long, as beeswax has a melting point of 64°C/147°F. Quickly dip the shallot back into the fluid wax and repeat the process until you've built up five layers.

Once the fifth layer has dried, remove the skewer and dip the root end into the wax to seal the hole and coat the entire shallot. Allow the wax to dry one last time, then set the shallot aside on a tray. Repeat the process until you've coated all the shallots.

Carefully place the wax-coated shallots on a tray in the fermentation chamber and age them for 8 to 10 weeks. It's important to ensure that the holding temperature of your fermentation chamber is both accurate and stable. This is where the heating mats and temperature controllers we discuss in the primer chapter can be quite useful (see "Building a Fermentation Chamber," page 42). If you're using such a rig, try to keep the heat source away from the shallots, so there are no hot spots that could potentially melt the wax.

As an example, if you've made your fermentation chamber out of a decommissioned chest freezer, offset the heating supply to one side and place the shallots on a tray on the other side, raised off the freezer bottom. A temperature controller will keep the temperature in the freezer steady to within a degree and allow the beeswax to remain solid. For even better results, place a small fan in the chest freezer as well to allow for better air circulation.

When it comes time to harvest, allow the shallots to cool to room temperature, then slice the wax open with a knife. They'll have the texture of a whole roasted shallot that you might find at the bottom of a pan of roast chicken. The shallots can also be stored in the wax in the fridge or freezer until you're ready to use them.

Suggested Use

Onion Soup

Thinking of black shallots as the most deeply caramelized onions imaginable will help you cook more creatively with them. And what's one of the first things that comes to mind when you think of caramelized onions? Onion soup. With a sharp knife, julienne 250 grams black shallots and cook them with enough rich beef stock to cover, along with a splash of fortified wine. Adjust the seasoning with salt and black pepper before transferring the broth into soup bowls and topping them with slabs of crusty bread and lots of grated Swiss cheese. Broil until the cheese is browned and stringy. When your friends and family marvel at the soup, tell them you started cooking it two and a half months ago.

Thinly slice black shallots for use in soup (or anywhere else you'd use caramelized onions).

439

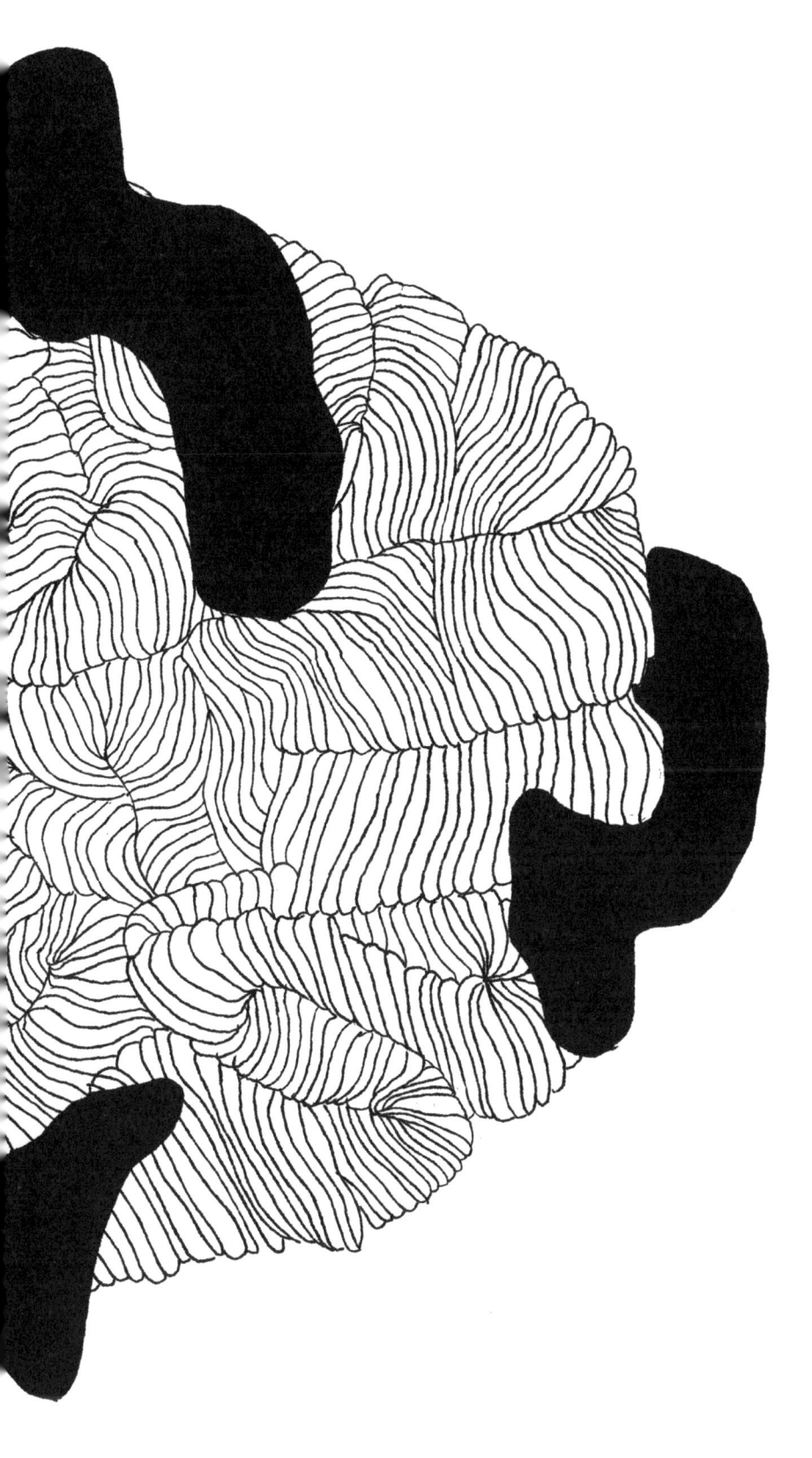

Equipment

There are a multitude of different ways to go about fermentation, and thus there is no single set of "correct" equipment. Some fermenters will swear by their grandmother's heirloom ceramic crock, while others make do with repurposed pickle jars. We'll stop short of suggesting specific items, lest we make you feel like you can't carry out a fermentation project unless you have exactly what we have. But here's some basic information for reference, should you be wondering what to buy online or from your local fermentation store.

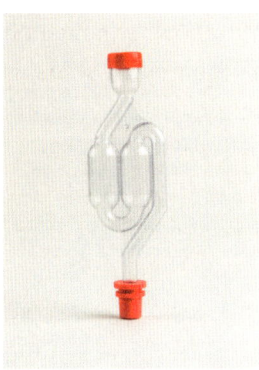

Airlocks

An indispensable tool when brewing alcohol, and useful for lacto-fermentation as well, an airlock consists of a water-filled S-shaped vent plugged into a rubber stopper. Air cannot enter through the airlock, but as microbes create gas inside the fermentation vessel, pressure builds until it is expelled through the vent.

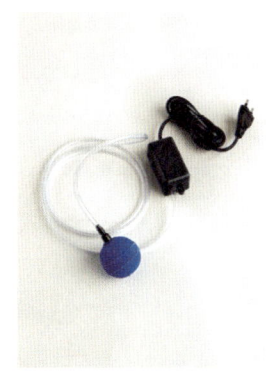

Air Pump and Air Stone

Common devices used to aerate vinegars and supply oxygen to the aerobic bacteria fermenting it, air pumps can be found at both pet stores and home-brew shops. Many come with stone aerators (air stones), but if you notice them deteriorating over time, you can also find metal aerators.

Ceramic Crocks

Ceramic crocks are tried and true vessels for fermented foods. They're opaque, so light can't enter (see sidebar, page 72), which prevents UV rays from damaging microbial cells. Many crocks come with the added feature of a small moat in which the lid sits. When filled with water, it acts as an airlock.

Cider Press

Normally used to press the mash of fermented fruit through a muslin bag, a cider press can also be used to harvest shoyu or the juice of lacto-fermented fruits and vegetables.

Dehydrator

This countertop appliance blows warm air over trays of ingredients, drying them slowly. Dehydrators are great for making use of a range of fermented goods, from lacto-fermented fruits to black vegetables.

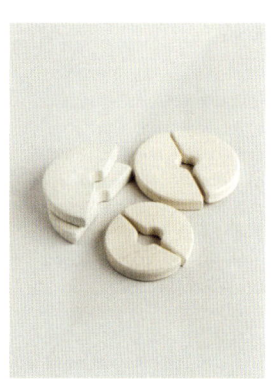

Fermentation Weights

Often included with fermenting crocks, glass or ceramic fermentation weights are used to press ingredients down beneath a water line and out of contact with the air. They're especially useful for making miso, and when lacto-fermenting sturdier products.

Glass Jars (Different Shapes and Sizes)

Glass is a great medium for fermenting kombuchas and lacto-ferments because it is inert and allows you to track your progress visually (see sidebar, page 72). Swing-top or screw-on lids are equally acceptable.

Grain Polisher

Used in Japan to remove the endocarp (outer layer) of rice by whirring the grains against an abrasive surface, making it easier for koji to take hold. A grain polisher will work equally well with lesser-known grains, like emmer or konini.

Humidifier

A small humidifier is necessary for incubating koji. These come in many shapes and types. Some operate with an ultrasonic plate that creates an ultrafine mist, while others disperse water into the air through evaporation. Either will work, but in general, smaller is better.

Juicer

Indispensable for extracting fruit and vegetable juices for vinegars and kombuchas, juicers come in many shapes and styles.

Kioke

Flared, open-top barrels traditionally used in Japan to ferment sake, miso, and shoyu, kioke are usually made of cedar, which imparts a distinct flavor to the ferment. The open top allows the mixture to be stirred or pressed down with weights.

Koji Shaker

You're looking for an icing/powdered sugar shaker: a simple metal cylinder outfitted with a wire screen lid. A shaker allows you to distribute *Aspergillus* spores onto trays of steamed rice or barley.

Muslin or Cheesecloth

Useful for straining mashes and ice-clarifying broths and stocks, these are made of either vinyl or cotton and can be reused with a thorough washing.

Nylon Mesh Strainer

Also called a fish net, this is a wire or plastic ring fitted with fine nylon mesh used to strain liquids and sift flours. A fine-mesh sieve lined with cheesecloth serves the same purpose.

Perforated Steel Tray

A perforated stainless-steel tray allows koji to grow on a sanitary surface with access to oxygen.

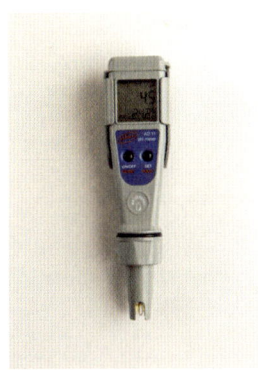

pH Meter

This handheld tool provides an accurate, digital measurement of a liquid's pH.

pH Test Strips

These chemically reactive strips of paper change color depending on the pH of a solution. You dip the strips into a liquid, then visually compare the color to a key.

Plastic Bucket

Food-grade plastic buckets are ideal vessels for larger batches of ferments like kombucha, miso, garum, vinegar, and alcohol (when equipped with an airlock). Note that plastic does have a tendency to absorb flavors, so it's best not to use the same bucket for different purposes.

Pot and Steamer

A simple three-piece setup (pot, lid, and perforated steamer insert) is used to cook grains.

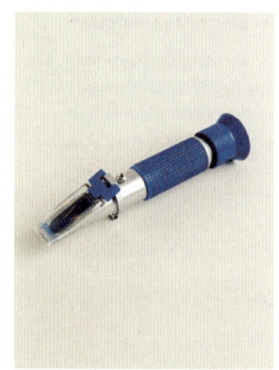

Refractometer

An optional but extremely handy tool, a refractometer allows you to determine the sugar content of a liquid by measuring the refraction of light. (The more sugar dissolved in water, the greater the refractive index.)

Rice Cooker

A large rice cooker set to "keep warm" can serve as a fermentation chamber for products that need to be held at higher temperatures, such as garums and black fruits and vegetables. Be sure to find a rice cooker that does not come equipped with an auto-off feature.

Space Heater or Heating Mat

Either of these devices can be used to heat a well-insulated fermentation chamber. (Heating mats—normally used to germinate seeds or heat reptile terrariums—are better suited for smaller chambers.) While many heaters are equipped with manual rheostats, they're best when combined with a temperature controller.

Styrofoam Cooler

Waterproof, well-insulated, cheap, and easily cleaned, Styrofoam coolers make ideal fermentation chambers. See page 47 for instructions.

Swing-Top Glass Bottles

Perfect for storing finished liquid ferments, such as vinegar and kombucha, because their seals are airtight, bottles can also be used to carbonate kombucha: Fill the bottles and leave them in the fridge for 1 to 2 weeks before consuming.

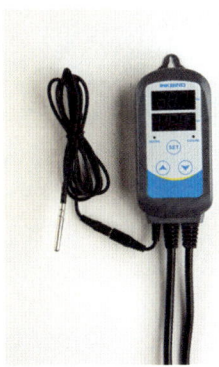

Temperature Controller

An electronic device for regulating the temperature of a fermentation chamber, it effectively functions as a thermostat that can be programmed to control a heat source of your choosing. Some models can also switch over to a cooling function should the temperature get too hot.

Vacuum Sealer and Vacuum Bags

A countertop vacuum sealer is extremely useful for lacto-fermentation and storage of all your ferments once complete. Clear vacuum bags allow you to see what's happening during fermentation and can easily be cut open to vent gas and resealed.

Wooden Barrel

Barrels impart unique flavors and allow for very slow evaporation of the liquids they hold. Alcohols and vinegars are the most common barrel-aged ferments, but garums and shoyus also benefit from aging in wood.

Wooden Koji Tray

Typically made from untreated planks of cedar, koji trays are the traditional vessel for growing *Aspergillus oryzae* on rice or barley. If all goes well, there's no need to wash the trays between batches of koji, as the fungus comes to live in the trays themselves.

447

Sources

An organic farmers' market or grocery store is your best source for the raw ingredients called for in this book. For fermentation-specific items, a well-stocked home-brew or fermentation shop will have almost everything you need. Beyond that, there are some things you can't find online. Here are a few sources we've used to procure specialized products.

Bee Pollen
bee-pollen.co.uk
911honey.com
rawliving.eu

Brewer's Yeast
hopt-shop.dk
themaltmiller.co.uk
yeastman.com

Defatted or Low-Fat Hazelnut Meal
bobsredmill.com
oelmanufaktur-rilli.de
paleo-paradies.de

Grasshoppers and Wax Moth Larvae
delibugs.nl
speedyworm.com
topinsect.net

Koji Spores
akita-konno.co.jp
americanbrewmaster.com
gemcultures.com
organic-cultures.com

Kombucha SCOBY
culturesforhealth.com
fairment.de
happykombucha.co.uk
hjemmeriet.com
kombuchakamp.com

Acknowledgments

For all the iterations this book has taken on its winding path to becoming what it is, both authors would like to thank the many individuals who either helped it along its way or afforded us the time to write it, among them, Thomas Frebel, Mette Brink Soberg, Benjamin Paul Ing, Junichi Takahashi, Jason White, Matt Orlando, Paula Troxler, Evan Sung, Jason Loucas, Laura and Andreja Lajh, Lizzie Ellison, Aralyn Beaumont, Paul D'Avino, Diego Gutierrez, Phil Hickman, Alex Petrician, Adriano Bruzzese, Anne Catherine Preißer, Priyanca Patel, Fiona Strouts, and Alessio Marcato, in addition to the entire Noma family, everyone at Atelier Dyakova, and the entire team at Artisan Books.

We would also like to thank authors whose amazing bodies of work inspired us to delve deeper into the practice, history, and science of fermentation at all. William Shurtleff and Akiko Aoyagi, Harold McGee and Sandor Katz.

Not least, very special thanks are in order to the team of amazing editors who helped shape the words of this book into a volume that is as enjoyable to read as it was to write: Chris Ying, Martha Holmberg, Dr. Arielle Johnson, and our publisher, Lia Ronnen.

RR & DZ

Index

A

acetic acid bacteria (AAB), 30, 33–34, 112, 113, 161, 163–69
acids, 59
Ajinomoto Group, 213, 335
Amazake, Sparkling Citric Koji, 248–51
apples
    Apple Kombucha, 136–39
    Black Apples, 424–27
ascorbic acid, 59
Asparagus, Lacto White, 92–95
Aspergillus luchuensis, 32, 220
    Citric Barley Koji, 242–45
Aspergillus oryzae, 26–27, 32, 40, 212, 213, 215–21, 223, 317, 330–32
    harvesting your own spores, 241
    in miso, 271
    variants of, 220–21
    see also koji
autolysis, 362, 366–67, 369

B

backslopping, 33–34
Bacon-Cep Vinaigrette, 85
bacteria, 19, 29, 30, 36
    acetic acid (AAB), 30, 33–34, 112, 113, 161, 163–69
    Clostridium botulinum, 38–39, 61
    Escherichia coli, 39
    lactic acid (LAB), 30, 33, 56–57, 59–62, 65
    Salmonella, 39, 41
    SCOBY, 34, 40, 111–18
baker's percentages, 41–42
balsamic vinegar, 168–69
    Black Garlic Balsamic, 206–9
    Elderberry Wine Balsamic, 200–205
barley, 216, 217
    Citric Barley Koji, 242–45
    Pearl Barley Koji, 230–41
    Roasted Koji "Mole," 262–64
    Sweet Citric Koji Water, 246–47
BBQ Sauce, Kombucha, 150
Béarnaise, Pear, 180

beef
    Beef Garum, 372–79
    burgers, 379
    Garlic-Peaso Grilled Beef, 300
Bee Pollen Garum, 396–99
berries, 59–60
    Berry-Rose Coulis, 135
    Breadso-Buttered Toast with Berries and Cream, 323
    Dressing for Summer Fruits, 304, 305
    Lacto Blueberries, 96–99
    Lacto Green Gooseberries, 104–7
beta-amylase, 32
beverages
    Black Hazelnut Milk, 435
    Gin and Rose Cocktail, 134
    Not Chocolate, 264
    Quatre Épices Cocktail, 150, 151
    Sparkling Citric Koji Amazake, 248–51
    Sweet Citric Koji Water, 246–47
    see also kombucha
black fruits and vegetables, 405–13
    Black Apples, 424–27
    Black Chestnuts, 428–31
    Black Garlic, 416–23
    Black Hazelnuts, 423–35
    moisture content and retention in, 412
    pungency in, 413
    sugar in, 413
    Waxed Black Shallots, 436–39
Blueberries, Lacto, 96–99
botulism, 38–39, 61
Brandied Black Apples, 427
Breadso, 320–23
breakfast
    Kombucha Syrup, 131
    Lacto Blueberries, 96–99
Brettanomyces, 31
brine, 61–62, 65–66
Brix, Adolf, 118
Brix scale, 115, 118
bruschetta, 90
broth
    Beef Garum, 379
    Black Garlic Skin Broth, 422
    Overnight Chicken Broth, 358
    Ramen Broth, 390

browning, 405–6, 409
butter
    Caramelized Rose Butter, 304
    Cep Shoyu Beurre Blanc, 354
    Grasshopper Butter, 394
    Lacto Koji Butter Sauce, 261
    Peaso Butter, 300, 301
    Shio Koji Butter, 266
butterfly effect, 278
buttermilk
    Buttermilk-Gooseberry
        Dressing, 107
    Shoyu-Buttermilk Fried Chicken,
        346
Butterscotch, Coffee Shoyu, 358

C

cabbage, 51
    Christmas Cabbage, 185
    Garlic-Peaso Cabbage, 299
    kimchi, 11, 50–51
    sauerkraut, 50–51, 60
calcium hydroxide, 315
Caramel, Shoyu, 347
caramelization, 406, 413
    Caramelized Rose Butter, 304
Carrots, Slow-Cooked, 192–93
Cashews, Roasted, 391
Celery Vinegar, 186–89
cep mushrooms
    Cep Shoyu, 352–55
    Cep Shoyu-Glazed Ceps,
        354, 355
    Lacto Cep Mushrooms,
        82–85
    Ryeso Tamari and Ryeso-
        Mushroom Glaze, 310, 311
Chang, David, 11
chaos theory, 278
Cheese, Fresh, Celery-Herb
    Vinegar with, 189
Chestnuts, Black, 428–31
chicken
    Overnight Chicken Broth,
        358
    Roasted Chicken Wing Garum,
        388–91
    Shoyu-Buttermilk Fried Chicken,
        346

Chocolate-Covered Black Apples,
    427
citric acid, 32, 59
Citric Barley Koji, 242–45
    Sparkling Citric Koji Amazake,
        248–51
    Sweet Citric Koji Water, 246–47
Clarke, Arthur C., 212
cleanliness and safety, 36–40
Clostridium botulinum, 38–39, 61
coffee
    Coffee Kombucha, 144–47
    Coffee Shoyu, 356–59
Confit, Koji Oil, 255
containers, glass versus ceramic,
    72
corn
    Maizo, 312–14
    Masa, 315
    on the cob, 99
Coulis, Berry-Rose, 135
cream
    Elderflower Crème Fraîche, 142,
        143
    Ryeso Cream, 309
Croutons, Crunchy Koji, 241
Crudités, 387
cucumbers
    Cucumber Soup, 189
    pickles, 58
Cure, Koji (Shio Koji), 265–67
Custard, Plum, 80, 81

D

desserts
    Berry-Rose Coulis, 135
    Black Garlic Ice Cream, 422
    Brandied Black Apples, 427
    Breadso-Buttered Toast with
        Berries and Cream, 323
    Candied Cep Mignardises, 85
    Chocolate-Covered Black
        Apples, 427
    Coffee-Kombucha Tiramisu,
        146
    Coffee Shoyu Butterscotch, 358
    Koji "Marzipan," 256
    Kombucha Syrup, 131
    Plum Custard, 80, 81

Pumpkin Seed Miso "Ice Cream,"
    327
Ryeso Cream, 309
Shoyu Caramel, 347
dressings
    Beef Garum Emulsion, 379
    Buttermilk-Gooseberry
        Dressing, 107
    Dressing for Summer Fruits,
        304, 305
Dryad's Saddle Shoyu, 337, 348–51
Dzikilpak, 325, 326–27

E

Egg Yolk Sauce, 378, 379
El Bulli, 13
Elderberry Wine Balsamic, 200–205
elderflower
    Elderberry Wine Balsamic,
        200–205
    Elderflower Kombucha, 140–43
enzymes, 26, 32, 65, 330
    browning and, 406
    protein-dismantling, 32, 366–67,
        368, 371
equipment, 442–47
Escherichia coli, 39
ethanol (ethyl alcohol), 166, 189

F

fermentation, 9–17, 19–23
    backslopping and, 33–34
    by-products from, 67
    cleanliness and safety in, 36–40
    defined, 26
    equipment for, 442–47
    experimenting with, 50–51
    microbes in, see microbes
    pH in, 36, 40–41, 125–26
    primary and secondary, 26–27
    primer on, 25–53
    rot versus, 29
    salt in, see salt
    seasonings and, 65–66
    store-bought ferments, 51–52
    taste and, 27–28
    wild, 33, 57

451

fish
fish sauce, 362, 364–65
see also seafood
fermentation chamber, 42, 223–27
covered speed rack, 43–45
Styrofoam, 47–49, 224–27
Flour, Koji, 252–57
Frebel, Thomas, 362
fruits
choosing, 62–65
see also berries
fungi, 29, 31–32
see also molds; yeasts

G

Gammel Dansk Vinegar, 198–99
garlic, 118, 412–13
Black Garlic, 416–23
Black Garlic Balsamic, 206–9
Garlic-Peaso Cabbage, 299
Garlic-Peaso Grilled Beef, 300
Roasted-Garlic Oil, 299
garum, 26–28, 331, 337, 362–71
autolysis in, 362, 366–67, 369
Beef Garum, 372–79
Bee Pollen Garum, 396–99
Grasshopper Garum, 392–95
history of, 364–65
Roasted Chicken Wing Garum, 388–91
Rose and Shrimp Garum, 380–83
salt in, 366–68
Squid Garum, 384–87
water in, 367–68
Yeast Garum, 400–401
Gazpacho, Mango, 154
gherkins, 94
Gibbons, Euell, 23
Gin and Rose Cocktail, 134
Giusti, Dan, 14
glutamate, 368–69
monosodium (MSG), 213, 368–69
glutamic acid, 27–28, 332, 368
gooseberries, 9, 10
Lacto Green Gooseberries, 104–7
grains, 215–18
Grasshopper Garum, 392–95

H

hand taste, 278
hazelnuts
Black Hazelnuts, 423–35
Hazelnut Miso, 316–19
herbs
Apple Kombucha Herb Tonic, 138, 139
Celery-Herb Vinegar with Fresh Cheese, 189
History of Japan, The (Kaempfer), 331
Hollandaise, Pear, 180
Honey, Lacto Mango-Scented, 100–103
humin, 335
Hummus, Smoked, 401
hydrolized vegetable protein (HVP), 335

I

Ice Cream, Black Garlic, 422, 423
"Ice Cream," Pumpkin Seed Miso, 327
Ikeda, Kikunae, 335, 369
intricity, 363

J

jiangs, 274–75, 330–31, 334, 363
Johannson, Patrick, 57
Johnson, Arielle, 13, 14

K

Kaempfer, Engelbert, 331
Keller, Thomas, 66, 135, 300
kecap manis, 334
Kim, Scott, 407
kimchi, 11, 50–51
kioke, 277, 333
koji, 26–28, 33, 212–27, 284, 331–32, 368–71
age of, in peaso, 283–84
Aspergillus luchuensis in, 32, 220, 243–44
Aspergillus oryzae in, *see* *Aspergillus oryzae*
Citric Barley Koji, 242–45
Dried Koji and Koji Flour, 252–57
Dryad's Saddle and Roasted Koji Sauce, 351
grains in, 215–18
harvesting your own spores, 241
Koji Cure (Shio Koji), 265–67
Lacto Koji Water, 258–61
in miso and peaso, 271, 283–84
Pearl Barley Koji, 230–41
Roasted Koji "Mole," 262–64
Sparkling Citric Koji Amazake, 248–51
Sweet Citric Koji Water, 246–47
koji muro, 223, 224
kombucha, 33, 34, 110–18
Apple Kombucha, 136–39
basic process for, 114
bottling, 126, 127
Coffee Kombucha, 144–47
Elderflower Kombucha, 140–43
Lemon Verbena Kombucha, 122–31
Mango Kombucha, 152–55
Maple Kombucha, 148–51
Rose Kombucha, 132–35
SCOBY in, 34, 40, 111–18
sugar in, 114–15
timing for, 115

L

lactic acid bacteria (LAB), 30, 33, 56–57, 59–62, 65
lacto-fermented fruits and vegetables, 9, 10, 33, 50–51, 56–67
brine in, 61–62, 65–66
by-products from, 67
choosing fruits and vegetables for, 62–65
cucumber pickles, 58
Lacto Blueberries, 96–99
Lacto Cep Mushrooms, 82–85
Lacto Green Gooseberries, 104–7

Lacto Koji Water, 258–61
Lacto Mango-Scented Honey, 100–103
Lacto Plums, 68–81
Lacto Tomato Water, 86–91
Lacto White Asparagus, 92–95
removing air in, 60
salt in, 61–62
seasonings and, 65–66
temperature for, 65
timing for, 66, 67
Lavoisier, Antoine, 163, 407
leathers
Black Apple Leather, 427
Chewy, Dried Lacto Plums, 79
Lacto Tomato Leather, 90
Leche de Tigre, 106–7
Lemongrass-Mango Vinaigrette, 155
Lemon Verbena Kombucha, 122–31
Lettuces, Grilled, 327
Locke, John, 331
Lorenz, Edward Norton, 277–78

M

MAD Symposium, 10–11
Maggi, Julius, 335
Maillard, Louis Camille, 409
Maillard reaction, 405–6, 409–11, 413
Maizo, 312–14
malic acid, 59
mango
Lacto Mango-Scented Honey, 100–103
Mango Kombucha, 152–55
Maple Kombucha, 148–51
marinades
Koji Cure (Shio Koji), 265–67
Marinade for Roasted or Grilled Meat, 184
"Marzipan," Koji, 256
Masa, 315
Mayonnaise, Koji, 255
measurements, see weights and measures
meats
Beef Garum, 372–79
burgers, 379

Garlic-Peaso Grilled Beef, 300
Koji Cure (Shio Koji), 265–67
Kombucha BBQ Sauce, 150
Lacto Plum Powder, 80
Plum Vinegar Marinade for Roasted or Grilled Meat, 184
meju, 276
melanins, 406–7
mercury oxide, 407–9
Micheli, Pier Antonio, 216
microbes (microorganisms), 19, 26, 28–32, 65
backslopping and, 33–34
cleanliness and safety in working with, 36–40
terroir and, 11
see also bacteria; fungi; molds; yeasts
Mignardises, Candied Cep, 85
Mignonette, Lacto-Plum Juice, 80
misos and peasos, 270–78, 330–32
Breadso, 320–23
environmental humidity and, 283
flavorings in, 285
floral, 303
Hazelnut Miso, 316–19
history of miso, 271–77
koji in, 271, 283–84
Koji-Miso Soup, 254
Maizo, 312–14
moisture content of, 280–83
pressure and air exposure in, 284–85
Pumpkin Seed Miso, 324–27
Rose Peaso, 302–5
Ryeso, 306–11
salt in, 279–80
soy in, 271–76
steps in making, 279
temperature and time in, 284
Yellow Peaso, 288–301
molds, 19, 31
Aspergillus luchuensis, 32, 220, 243–44
Aspergillus oryzae, see Aspergillus oryzae
pathogenic, 39–40, 60
"Mole," Roasted Koji, 262–64
moromi, 332, 333

MSG (monosodium glutamate), 213, 368–69
mushrooms
Cep Shoyu, 352–55
Cep Shoyu-Glazed Ceps, 354, 355
Dryad's Saddle Shoyu, 337, 348–51
Lacto Cep Mushrooms, 82–85
Ryeso Tamari and Ryeso-Mushroom Glaze, 310, 311

N

Noma, 8–17, 19–23
Abalone Schnitzel and Bush Condiments, 227
Bergamot Kombucha with Native Mint, 113
Berries and Greens Soaked in Vinegar for One Year, 159
Chilled Oysters and Salted Green Gooseberries, 23
Chocolate from Native Jaguar Cocoa and Mixe Chile, 65
Deep-Sea Snow Crab and Cured Egg Yolk, 371
Garlic Flower, 13
Just-Cooked Octopus with "Dzikilpak," 283
Langoustine and Douglas Fir, 272
Pear and Roasted Kelp Ice Cream, 411
Roasted Bone Marrow, 21
Sea Snail Broth, 17
Soft-Boiled Egg and Black Garlic, 405
Sweet Peas, Milk Curd, and Sliced Kelp, 62
Unripe Macadamia Nuts and Spanner Crab, 215
Nordic Food Lab, 9–10
nuts, 317
Black Chestnuts, 428–31
Black Hazelnuts, 423–35
Hazelnut Miso, 316–19
Roasted Cashews, 391

453

O

oils
    Cep-Oil Companion, 85
    Koji-Infused Oil, 255
    Pollen Oil, 398
    Roasted-Garlic Oil, 299
onions
    Onion Salad, 319
    Onion Soup, 439
    Pissaladière, 386, 387
Orlando, Matt, 411
oxidation, 407–9
oxygen, 60, 61

P

Paillieux, Nicolas-Auguste, 334
pancakes
    Kombucha Syrup, 131
    Savory Pancakes, 394
Parsnips Glazed with Coffee
        Kombucha, 146, 147
pasta
    dried lacto plums in, 79
    Pasta with Egg Yolk Sauce,
        378
    Stuffed Pasta, 430, 431
pathogens, 36–40, 60
peas
    fresh, risotto, 80
    yellow, see yellow peas
pears
    Honey-Poached Pears, 103
    Perry Vinegar, 172–81
peaso, see misos and peaso
pH (potential of hydrogen), 36,
        40–41, 125–26
phenols, 406
pickles, 65
    cucumber, 58
    Quick Pickles, 193
    Tomato-Water Pickles, 89
Pissaladière, 386, 387
plums
    Lacto Plums, 68–81
    Plum Vinegar, 182–85
    Rose-Plum Sauce for Duck, 134
porcini mushrooms, see cep
        mushrooms

Potatoes, Koji Mole-Glazed, 264
poultry
    Koji Cure (Shio Koji), 265–67
    Kombucha BBQ Sauce, 150
    Overnight Chicken Broth, 358
    Roasted Chicken Wing Garum,
        388–91
    Rose-Plum Sauce for Duck, 134
    Ryeso Tamari and Ryeso-
        Mushroom Glaze, 310
    Shoyu-Buttermilk Fried Chicken,
        346
Priestly, Joseph, 407, 409
proteins, 27, 28, 216, 218
Pumpkin Seed Miso, 324–27
pyrolysis, 406

Q

Quatre Épices Cocktail, 150, 151

R

ramson capers, 9
Ramen Broth, 390
raspberry juice, 60
Reade, Ben, 10
redox reactions, 407–9
refractometers, 118, 125, 126, 181
Relish Gooseberry, 106
rice, 216–17
    Bee Pollen Risotto, 399
    pea risotto, 80
Rittman, Roland, 9
rose
    Rose and Shrimp Garum,
        380–83
    Rose Kombucha, 132–35
    Rose Peaso, 302–5
Ryeso, 306–11
    Black Garlic Vinegar and Ryeso
        Sauce, 208, 209

S

Saccharomyces cerevisiae, 31
safety and cleanliness, 36–40
Salmonella, 39, 41

salt, 41–42
    brine, 61–62, 65–66
    in garum, 366–68
    lacto-fermentations and, 61–62
    measurements and, 36, 41–42
    in miso and peaso, 279–80
    type of, 42
sauerkraut, 50–51, 60
sauces
    Black Garlic Vinegar and Ryeso
        Sauce, 208, 209
    Breadso Sauce, 323
    Cep Shoyu Beurre Blanc, 354
    Coffee Shoyu Butterscotch,
        358
    Dryad's Saddle and Roasted Koji
        Sauce, 351
    Egg Yolk Sauce, 378, 379
    Lacto Koji Butter Sauce, 261
    Pear Hollandaise or Béarnaise,
        180
    Ryeso Cream, 309
    Sole à la Grenobloise, 434–35
    Tomato Sauce, 90
    Whiskey Vinegar Sauce, 197
SCOBY, 34, 40, 111–18
seafood
    Caramelized Rose Butter, 304
    Clams and Cockles, 251
    Fish Cure, 314
    fish en papillote, 247
    Fish Glaze, 358
    Koji-Breaded Fish, 256, 257
    Koji Cure (Shio Koji), 265–67
    Lacto-Plum Juice Mignonette,
        80
    Lacto Tomato Water, 86–91
    Rose and Shrimp Garum,
        380–83
    Sautéed Shrimp, 193
    Shoyu-Oyster Emulsion, 346
    Squid Garum, 384–87
seasonings, 65–66
    Lacto Blueberries, 96–99
    Leche de Tigre, 106–7
Shallots, Waxed Black, 436–39
Shio Koji (Koji Cure), 265–67
shoyu, 277, 330–37
    acid hydrolysis method for,
        334–35
    Cep Shoyu, 352–55

Coffee Shoyu, 356–59
Dryad's Saddle Shoyu, 337, 348–51
methodology for, 337
Nordic, 335–36
Yellow Pea Shoyu, 338–47
shrimp
Rose and Shrimp Garum, 380–83
Sautéed Shrimp, 193
Smoothie, Apple-Vegetable, 138
S'mores, 319
soups
Beef Garum, 379
Breadso Soup, 322
Butternut Squash Soup, 383
Cucumber Soup, 189
Koji-Blanched Vegetables, 254
Koji-Miso Soup, 254
Mango Gazpacho, 154
Onion Soup, 439
Pearl Barley Koji in, 241
soy, 271–76
soy sauce, see shoyu
specific gravity, 118, 181
spirit vinegars, 167
squash
Butternut Squash Soup, 383
Butternut Squash Vinegar, 190–93
Squid Garum, 384–87
Stalking the Wild Asparagus (Gibbons), 23
Stock, Koji, 254
store-bought ferments, 51–52
sugar
in blackening, 413
Brix scale and, 115, 118
in kombucha, 114–15
Lacto Mango-Scented Honey as replacement for, 103
symbiosis, 112
syrups
Kombucha Syrup, 131
Maple Kombucha Syrup, 150

T

tamari, 277, 330, 331, 335
Peaso Tamari Reduction, 298

Ryeso Tamari and Ryeso-Mushroom Glaze, 310, 311
taste, 27–28
umami, 28, 213, 216, 218, 335, 369
Tiramisu, Coffee-Kombucha, 146
tomatoes
Lacto Tomato Water, 86–91
Roasted Tomatoes, 399
Tonic, Apple Kombucha Herb, 138, 139
Tostadas, 315
tuong, 334

U

umami, 28, 213, 216, 218, 335, 369

V

vegetables
Apple-Vegetable Smoothie, 138
choosing, 62–65
Crudités, 387
Koji-Blanched Vegetables, 254
Mashed Root Vegetables, 304
Vildgaard, Torsten, 9, 10
vinegar, 33, 34, 113, 158–69
balsamic, 168–69
barrel aging of, 168, 204, 205
Black Garlic Balsamic, 206–9
Butternut Squash Vinegar, 190–93
Celery Vinegar, 186–89
Elderberry Wine Balsamic, 200–205
Gammel Dansk Vinegar, 198–99
history of, 162–63
Perry Vinegar, 172–81
Plum Vinegar, 182–85
spirit, 167
two-step process for, 164
Whiskey Vinegar, 194–97
vinaigrettes
Cep-Bacon Vinaigrette, 85
Lemon Verbena Kombucha Vinaigrette, 131

Mango-Lemongrass Vinaigrette, 155
Perry Vinaigrette, 180
vodka, 166, 189

W

water activity, 367–68
weights, fermentation, 74
weights and measures, 36, 52–53
baker's percentages, 41–42
salt and, 36, 41–42
Westh, Søren, 9
Whiskey Vinegar, 194–97
Williams, Lars, 13, 14
Wiuff, Søren, 93

Y

yeasts, 19, 31, 161
Brettanomyces, 31
in kombucha, 112, 113
Saccharomyces cerevisiae, 31
SCOBY, 34, 40, 111–18
Yeast Garum, 400–401
yellow peas
Yellow Pea Shoyu, 338–47
Yellow Peaso, 288–301
see also misos and peaso

Check out
    @nomaferments
on Instagram for inspiration,
and share your own
creations online with the
hashtag #nomaferments

Library of Congress Cataloging-in-Publication Data

Names: Redzepi, René, author. | Zilber, David (Chef)
Title: The Noma guide to fermentation / René Redzepi and David Zilber.
Description: New York : Artisan, a division of Workman Publishing Co., Inc. [2018] | Series: Foundations of flavor | Includes index.
Identifiers: LCCN 2018003633 | ISBN 9781579657185 (hardcover : alk. paper)
Subjects: LCSH: Fermentation—Biotechnology. | Flavor. | Fermented foods. | Noma (Restaurant : Copenhagen, Denmark)
Classification: LCC TP371.44 .R43 2018 | DDC 664/.024—dc23
LC record available at https://lccn.loc.gov/2018003633

Published by Artisan,
an imprint of Workman Publishing,
a division of Hachette Book Group, Inc.
1290 Avenue of the Americas
New York, NY 10104
artisanbooks.com

The Artisan name and logo are registered trademarks of Hachette Book Group, Inc.

Printed in China (APO) on responsibly sourced paper

20 19 18 17 16 15 14 13 12